Birkhäuser

Mathematics and the Built Environment
Volume 7

Throughout history a rich and complex relationship has developed between mathematics and the various disciplines that design, analyse, construct and maintain the built environment.

This book series seeks to highlight the multifaceted connections between the disciplines of mathematics and architecture, through the publication of monographs that develop classical and contemporary mathematical themes – geometry, algebra, calculation, modelling. These themes may be expanded in architecture of any era, culture or style, from Ancient Greek and Rome, through the Renaissance and Baroque, to Modernism and computational and parametric design. Selected aspects of urban design, architectural conservation and engineering design that are relevant for architecture may also be included in the series.

Regardless of whether books in this series are focused on specific architectural or mathematical themes, the intention is to support detailed and rigorous explorations of the history, theory and design of the mathematical aspects of built environment.

Alberto Pugnale · Alberto Bologna

Architecture Beyond the Cupola

Inventions and Designs of Dante Bini

Birkhäuser

Alberto Pugnale
Faculty of Architecture, Building and Planning
The University of Melbourne
Melbourne, VIC, Australia

Alberto Bologna
Facoltà di Architettura, Dipartimento di Architettura e Progetto (DiAP)
Sapienza Università di Roma
Rome, Italy

ISSN 2512-157X ISSN 2512-1561 (electronic)
Mathematics and the Built Environment
ISBN 978-3-031-26734-5 ISBN 978-3-031-26735-2 (eBook)
https://doi.org/10.1007/978-3-031-26735-2

This book is published under the imprint Birkhäuser, www.birkhauser-science.com by the registered company Springer Nature Switzerland AG
The registered company address is: Gewerbestrasse 11, 6330 Cham, Switzerland

Preface

Over the last few decades, the impact of digital design processes and fabrication techniques on architecture has grown in an exponential manner, as demonstrated by Greg Lynn's amoebic "BLOBs" and Lars Spuybroek's "free-forms". According to Lynn, designing BLOBs is closely linked to the invention of new construction systems: it is a process that requires a constant redefinition of the relationship between architectural form and its construction. In free forms, the adjective "free" indicates the freedom to create architectural forms, irrespective of any structural or construction principle: it is an approach that has been taken to an extreme in the purely virtual "trans-architecture" of Marcos Novak. The development of digital fabrication has also challenged traditional serial production: through a direct "file-to-factory" process, a single parametric model can now be materialised in a multitude of unique spatial variations, without any significant increase in construction costs. These are only a few of the themes covered in the "Non Standard Architectures" exhibition, that was held at the Centre Pompidou in Paris between December 2003 and March 2004. From the name of this exhibition, it is possible to gather that the curators did not entirely solve the challenging task of theoretically and historically classifying the heterogeneous group of projects on show: they decided to simply acknowledge the "non-standard" nature of such works.

In such a complex scenario, this book presents a unique approach to the design of thin concrete shells which is based on the development of an innovative and automated construction technique to generate architectural and structural form: the Binishell. The architect Dante Bini invented this singular construction system in 1964 in Italy. The concept involves casting a reinforced concrete structure on the ground, and this structure is then gradually raised by means of air: tonnes of fresh cast concrete take on a final structural shape thanks to the use of a special pneumatic formwork and a self-shaping metallic reinforcement. The Binishell is an invention that falls perfectly within a specific design culture of the 1960s, in which the desire to explore new architectural forms went hand in hand with the trend and popularity of thin concrete structures.

The book focuses on the architectural design challenges and compositional implications of designing shell structures by means of automated construction techniques.

The purpose is to disseminate Dante Bini's inventions, designs and built projects, while contextualising his work within the cultural period in which he grew up. At the same time, the ambition of the authors has been to explore the past to better understand the current trends and developments in digital design and fabrication, to inspire and inform the potential future uses of innovative construction automation techniques and the next generation of new architectural forms and languages.

Chapter 1 is dedicated to the invention of the Binishell. It sheds light on the multifaceted figure of the Italian architect Dante Bini, who is an inventor, a designer, and an adventurous entrepreneur. This chapter also provides the context in which Bini developed his construction systems and illustrates the experimental and empirical nature of a singular design culture of the 1960s.

Chapter 2 constructs a theoretical narrative through which the Binishell can be interpreted as a new archetypal hut in *structural art*, a concept originally conceived by Ada Louise Huxtable in 1960 and then made popular by David P. Billington in 1983. The archetypal hut can be defined as a natural and pragmatic response to the human need for shelter: the perfect synthesis between architectural and structural form and its construction. Similarly, the Binishell system can also be seen as a very rational response to the need to build more efficient and economical buildings that require fewer resources during their construction and throughout their lifespan. By offering an interpretation of the terminological dichotomy that exists between *dome* and *cupola* in the context of *structural art*, this chapter also investigates and discusses the spatial and tectonic qualities of the Binishell, as well as of the other systems that have been derived from this invention.

Chapter 3 focuses on the design of dwellings that feature a circular plan. It constructs a narrative that highlights how the Binishell system has evolved over the years to better respond to the challenging task of organising human life inside a space with no orthogonal elements or conventional doors and windows. In other words, this chapter describes how the limits of the original Binishell patent became a source of inspiration for Bini and led to the development of other automated construction systems, such as the Minishell and the Pack-Home. This body of work also inspired Bini's son—Nicolò—to redesign a fluid and non-modular contemporary version of the Binishell system with the idea of investing in construction innovation to address the current sustainability issues of the built environment.

Chapter 4 investigates the issue of cutting the openings in a Binishell structure and, more broadly speaking, the architectural implications of trimming and puncturing a concrete shell in different ways to let people and natural light enter. The chapter begins with a brief description of traditional form-finding techniques to highlight that it is only the pneumatic or "inflated hill" method, invented by Heinz Isler in 1954, that does not automatically create openings as part of the process of structural form generation. Like Isler's method, the Binishell uses an airtight pneumatic membrane to find structural form. The critical difference between the two is that a Binishell finds form and erects a concrete shell simultaneously, thereby creating an impenetrable fortress that literally has to be "broken" to be accessed and used. The core of the chapter concentrates on the compositional aspects of designing openings in Binishell structures. The narrative draws inspiration from Ross Styles' reflections on Bini's work, in which shell openings are clustered into two categories: those that are created at the base of the structure, such as the openings of the cupola of the Hagia Sophia Grand Mosque, and those designed at the apex of the dome, for example, the oculus of the Pantheon in Rome. The final part of the chapter describes the construction systems that were derived from the Binishell, that is, the Minishell and the recent System A developed by Nicolò Bini, in which the openings are automatically generated during the construction process.

Chapter 5 focuses on the concepts of modularity and standardisation in relation to the design of Binishells. It begins by illustrating projects in which modular Binishell structures have been repeated to create assemblies of buildings that affect and interact with the design of the surrounding landscape. The geometric intersection of Binishells is then introduced to understand how larger-span structures and more complex architectural and structural bodies can be created using a modular system, as in the case of the Narrabeen North Public School or the Space City Shopping Centre in Australia. The chapter ends with a discussion on the geometric and structural variations and complexity made possible by recent parametric modelling tools and digital fabrication techniques. In this context, Nicolò Bini's systems and projects demonstrate how Dante Bini's legacy can be a powerful source of inspiration to define future non-modular design and construction innovation strategies.

Chapter 6 uses Nicolò Bini's explorations as a vehicle to discuss how the concept of construction innovation can unfold in a period in which free-form architecture and even purely virtual designs seem to take priority over tectonics and the material and performative aspects of buildings. This chapter primarily focuses on a villa in Malibu that Nicolò designed for Robert Downey Jr.: a building that represents the latest evolution of the Binishell system originally invented by Dante Bini in 1964. The aim is to discuss the complex relationship that exists between digital design and construction automation in a particularly fascinating period of contemporary architecture: the apparent anti-tectonic nature of Greg Lynn's inform BLOBs is more than ever linked to the invention of new construction and fabrication systems, thus strengthening the relationship between architectural form and its construction.

In summary, this book is about Dante Bini's inventions and designs. However, the arguments presented herein can easily be applied to the analysis and interpretation of a variety of other thin shell structures, as well as any other architectural project in which the relationship between form, structure and construction is pivotal.

We hope that, in time, this book will become a reference on the possible ways and methods of conducting research in architecture that are based on the critical analysis of precedents, particularly those in which the close relationship between form, structure and construction technique had been pivotal throughout the design process, to better understand the present and the future of the discipline and, more broadly speaking, the built environment.

Melbourne, Australia
Rome, Italy
2022

Alberto Pugnale
Alberto Bologna

Acknowledgements

This book is the result of a long study on Dante Bini's inventions and projects, which began in 2013, about a year after Alberto Pugnale moved to Australia to work as a Lecturer at The University of Melbourne. At that time, Pugnale was an early career researcher looking for financial support. In the process of applying for a small internal grant, he was invited by his former colleague, Marianna Nigra, to investigate an Italian architect and inventor, who had also emigrated to Australia, in the 1970s, to design and build a number of concrete domes through an innovative construction system: the Binishell. Thanks to her suggestion, Pugnale was awarded a small internal grant (Early Career Research Grant) by the Faculty of Architecture, Building and Planning (ABP) at The University of Melbourne. In July 2013, Alberto Pugnale and Alberto Bologna, who at that time was a post-doc scientist at the École Polytechnique Fédérale de Lausanne, decided to collaborate on this project, following a conversation that had begun during the Second International Conference on "Structure and Architecture" (ICSA) in Guimarães, Portugal. Having been fellow university students, and both being engaged in academic research on the relationship between form and structure in contemporary architecture and building engineering, Pugnale and Bologna decided to join forces and take advantage of their complementary scholar profiles to investigate Dante Bini's inventions and designs. This project has grown over the years and has led to the publication of this book, which has once again been supported by the Faculty of Architecture, Building and Planning (ABP) of The University of Melbourne with a Publication Grant, obtained in 2018, and by the Department of Architecture and Design (DiAP) of the Sapienza University of Rome for the acquisition of publication rights for some of the images that have been used.

We would like to thank profusely the editors of the book series, Kim Williams and Michael J. Ostwald. They supported this project in many different ways, from an initial invitation to publish our preliminary research findings in the *Nexus Network Journal* to the suggestion of writing an entire book on Dante Bini's work. Both editors have patiently reviewed our drafts and helped us fine tune the book structure, title and overall narrative.

Our biggest thanks goes to Dante Bini, who has generously shared drawings, images, memories and technical information with us, all of which have been essential to complete this book. Pugnale was kindly hosted by Bini in his house in St. Helena, California, in 2016, where he found a very welcoming environment and all the support he needed to study Bini's personal archive; Bologna visited Bini in Arezzo in 2022, where he had many constructive conversations with him that helped clarify some issues that had arisen during the final stages of the writing of the book. Bologna also promoted the organisation of Dante Bini's keynote lecture at the Sapienza University of Rome on 25 February 2022, which was an important occasion to discuss his work with Bini himself and with other scholars. At the moment of writing, the recording of the lecture (in Italian) is available on: https://www.youtube.com/watch?v=JgE7jKbjFPc&t=919s.

Another essential thank you goes to Nicolò Bini, who helped us frame certain issues related to his father's vast production and provided us with a complete set of drawings and construction photographs of his recent projects.

We are not the only academics who have studied Dante Bini's work in recent years. We are part of a friendly and motivated group—the "Bini gang"—that includes Carlo Dusi, William McLean and Giulia Ricci, who are also investigating the Binishell and the related inventions. Thanks are due to all of them for having shared their knowledge and archival material with us to allow this book to be completed and to further disseminate Bini's legacy.

Some concepts presented and discussed in this book have been developed from previously published conference papers and journal articles. Although such material was substantially revised, expanded and updated for this publication, it is worth noting that, in Chap. 1, the section on the roots of the Binishell system draws inspiration from: Pugnale, Alberto, and Alberto Bologna. 2014. Dante Bini's air structures (1964–1979). From early Italian prototypes to the Australian experience. In *Proceedings of the First Construction History Society Conference. Queen's College, University of Cambridge, 11–12th April 2014*, ed. James W. P. Campbell, Wendy Andrews, Nicholas Bill, Karey Draper, Patrick Fleming, Yiting Pan, 355–365. Cambridge UK: Construction History Society-Short Run Press. In Chap. 2, a preliminary discussion on the connection between the contents of Mario Salvadori's *Why buildings stand Up* and the design concepts expressed by Bini was initially developed and presented in: Pugnale, Alberto and Alberto Bologna. 2017. Dante Bini's form-resistant Binishells. *Nexus Network Journal. Architecture and Mathematics* 3: 681–699. Bini's emigration to Australia and some details of his Australian Binishells were initially discussed in: Pugnale, Alberto, and Alberto Bologna. 2015. Dante Bini's "new architectural formulae": construction, collapse and demolition of Binishells in Australia 1974–2015. In *Architecture. Institutions and Change. Proceedings of the Society of Architectural Historians Australia and New Zealand Vol. 32*, ed. Paul Hogben, Judith O'Callaghan, 488–499. Sydney: SAHANZ. The maturity of the concepts presented in this book is also the result of the work done by paper reviewers and editors, which is generally invisible to readers and often goes unnoticed.

Book Images

This book includes a large number of drawings and photographs, many of which have never been published before or have specifically been prepared for this publication. See below for details and figure credits:

Agovision (A), MAD Architects (MA):
Fig. 6.7
ArchExist (AE), MAD Architects (MA):
Fig. 6.8
Amanzio Farris (AF):
Figs. 1.31, 2.14
Alo Zanetta photographer (AZ) and Mario Botta architect (MB):
Fig. 4.30 (bottom)
Alberto Pugnale (AP):
Figs. 2.18, 3.1, 3.2, 3.5, 3.7, 3.9, 3.10, 3.11, 3.12, 3.13, 3.14, 3.17, 3.18, 3.19, 3.20, 3.21, 3.22, 3.23, 3.24, 3.25, 3.26, 3.27, 3.28, 3.29, 3.32, 3.33, 3.34, 4.2, 4.3, 4.4, 4.5, 4.6, 4.7, 4.8, 4.9, 4.11, 4.21, 4.22, 5.4, 5.5, 5.7, 5.8, 5.12, 5.13, 5.14, 5.17, 5.19, 5.21, 5.22, 5.23, 5.24, 5.25, 5.28
Carlo Dusi (CD):
Fig. 4.12
Collection Frac Centre-Val de Loire, Donation Pascal Häusermann, photographer Philippe Magnon (CFCVL):
Fig. 3.4
Dante Bini (DB):
Cover illustration, Figs. 1.6, 1.11, 1.12, 1.13, 1.14, 1.15, 1.16, 1.17, 1.18, 1.19, 1.20, 1.21, 1.22, 1.23, 1.24, 1.25, 1.28, 1.29, 2.1, 2.13, 2.15 (photographer Max Dupain), 2.21, 2.22, 2.23, 2.24, 2.25, 3.8, 3.15, 3.16, 3.30, 3.31, 4.1, 4.15, 4.16, 4.17, 4.18, 4.19, 4.23, 4.24, 4.25, 4.31, 4.32, 4.33, 4.34, 4.35, 4.36, 4.37, 4.38, 4.39, 4.40, 5.1, 5.2, 5.6, 5.9, 5.10, 5.11, 5.15, 5.18, 5.26, 5.27, 5.29, 5.30
Diana Lanciotti, www.dianalanciotti.it (DL):
Fig. 1.30
De Rebus Sardois (DRS):
Fig. 4.30 (top)
Ekkehard Ramm (ER):
Fig. 4.10
The Frank Lloyd Wright Foundation Archives (The Museum of Modern Art | Avery Architectural & Fine Arts Library, Columbia University, New York) (FLWFA):
Fig. 1.5
Atelier Frei Otto (FO):
Figs. 6.14, 6.15
Gregory Burgess Architects (GBA):
Fig. 3.2

Giulia Ricci (GR):
Figs. 1.26, 2.20
Hotel Ariston Molino Buja (HAMB):
Fig. 4.20
John M. Johansen (JMJ):
Fig. 6.9
Kawaguchi & Engineers (KE):
Fig. 5.20
Library of Congress, Prints and Photographs Division, Washington D.C., Paul Rudolph Collection, PR 13 CN 2001:126 (LC, PRC):
Fig. 5.3
Steve Kornher and Lloyd Turner (LT):
Fig. 3.6
Lurati Muttoni Partner (LMP):
Figs. 6.1, 6.2
Malvern Girls' College (MGC). The authors thank Manuel Cresciani for providing us with the original photograph:
Fig. 1.27
The Huntington Library, San Marino, California, photographer Maynard L. Parker (MLP, HLSM):
Fig. 1.4
The Museum of Modern Art, New York/Scala, Firenze (MoMA):
Figs. 2.2, 2.3, 2.4, 6.10
Nicolò Bini (NB):
Figs. 1.32, 5.31, 5.32, 5.33, 6.11, 6.12, 6.13, 6.16, 6.17, 6.18, 6.19, 6.20, 6.21, 6.22, 6.23, 6.24, 6.25, 6.26, 6.27, 6.28, 6.29, 6.30
OPEN Architecture (OA):
Figs. 6.4, 6.5, 6.6
Pino Appeddu for De Rebus Sardois (PA, DRS):
Fig. 4.14
Romain Courtemanche (RC):
Figs. 2.17, 4.13
Ross Styles (reproduced from his thesis):
Figs. 4.26, 4.27, 4.28, 4.29
Simone Mengani, photographer (SM). Lurati Muttoni Partner (LMP):
Fig. 6.3
Sergio Poretti (SP). The authors thank Tullia Iori for providing us with the original photograph:
Fig. 2.19
Vittorio Garatti (VG):
Fig. 5.16
William McLean (WML):
Fig. 1.1

Reasonable efforts have been made to trace the copyright holders of the visual material reproduced in this book. The publisher, Alberto Pugnale and Alberto Bologna apologise for any errors and omissions that should be brought to our attention and which will be corrected in future editions.

Authors' Contributions

Alberto Bologna and Alberto Pugnale conceived and designed the book structure together, and defined the topics that had to be covered in each chapter, the research methodologies, the critical-interpretative approach that had to be adopted and the overall narrative of the text and figures in the book.

Alberto Pugnale collected the archival documents on the Australian Binishells from the New South Wales State Archives and visited many of the remaining structures throughout the country, some of which have local archives. Pugnale also visited Dante Bini in St. Helena in 2016 to discuss his work and scan his personal archive. Alberto Bologna searched for and obtained the various patents filed by Dante Bini in Italy and other countries, and he visited various Italian Binishells.

Alberto Bologna wrote the initial drafts of Chaps. 1, 2 and 6, while Alberto Pugnale wrote the preliminary drafts of Chaps. 3, 4 and 5. Alberto Pugnale translated and edited parts of Chaps. 1, 2 and 6 from the Italian into English. Marguerite Jones also translated many parts of Chaps. 1, 2 and 6 from the Italian into English and patiently proofread the entire book. The final text is the outcome of the joint discussions, reviews and contributions of both authors. Pugnale and Bologna also worked together on the search for and selection of the final book images. All the recent CAD drawings of Binishell projects were prepared by Alberto Pugnale.

Contents

About the Authors

Alberto Pugnale is an architect and Senior Lecturer in Architectural Design at the Faculty of Architecture, Building and Planning at The University of Melbourne (Australia). He is a codirector of the Advanced Digital Design + Fabrication (ADD+F) Research Hub and a founding member of the Bioinspiration Hallmark Research Initiative. Alberto has a PhD in architecture and building design and an MSc in architecture and construction. In 2007, he won the IASS HANGAI Prize, an international competition on research papers related to the field of shell and spatial structures for young researchers under 30. In 2008, he won a research scholarship, granted by the ISI Foundation (Turin, Italy), related to the study of complex architectural and structural bodies. From 2010 to 2012, he was an Assistant Professor of Structures at Aalborg University (Denmark). He has been an invited lecturer in various countries, including China, France, Mexico, Italy, Switzerland and the United States. At present, he is a member of the *International Association for Shell and Spatial Structures* (IASS). He is also a reviewer for international journals and a member of the editorial board of the *Nexus Network Journal: Architecture and Mathematics* and the *International Journal of Space Structures*.

Alberto Bologna is an architect and a tenure-track Assistant Professor (RTDb) in Architectural Design at the Faculty of Architecture—Department of Architecture and Design (DiAP) at the Sapienza University of Rome (Italy). Alberto has a PhD in architectural history and an MSc in architecture and construction. His research is focused on design cultures that are rooted in construction principles and he investigates the relationship between form, structure, tectonics, ornament and spatial quality in contemporary architecture and engineering. From 2011 to 2015, he was a post-doc scientist at the EPFL–École Polytechnique Fédérale de Lausanne, in Switzerland. From 2017 to 2020, he was a fixed-term Assistant Professor (RTDa) in Architectural Design at the Politecnico di Torino (Italy). Between 2018 and 2019, he was a visiting scholar at Tsinghua University in Beijing (China), where he taught design studios on exposed concrete. Alberto also worked as an Adjunct Professor in Architectural Design and Theory at the University of Genoa (2015–16), the University of

Ferrara (2016–17), the Politecnico di Milano (2016–17 and 2020–21), the Politecnico di Torino (2015–17), and the SUPSI–University of Applied Sciences and Arts of Southern Switzerland (2020–21). Alberto has authored several books that investigated the relationship between architectural design and construction, including *Pier Luigi Nervi Negli Stati Uniti* (Firenze University Press, 2013), *The Rhetoric of Pier Luigi Nervi. Concrete and Ferrocement Forms* (with R. Gargiani, EPFL Press-Routledge, 2016), *The Resistance of Laugier. The Classicism of Murcutt* (LetteraVentidue, 2019) and *Chinese Brutalism Today. Concrete and Avant-Garde Architecture* (ORO Editions, 2019). He is a member of the editorial board of several international peer-reviewed journals and a correspondent for the Swiss architectural magazine *Archi*.

Chapter 1
The Invention of the Binishell

1.1 The Man Who Built Domes

The Binishell system is a construction technique that was invented by Dante Bini in 1964. This system makes it possible to give shape to an architectural concept generated by the implementation of such a visionary technique that it can be considered as totally utopian. After almost sixty years from the first application of a Binishell, it is still capable of surprising and leaving young engineers and architects who come across photographs or videos of its construction incredulous.

The concept consists in the erecting of thin reinforced concrete shells cast on the ground and then gradually raised by means of a pneumatic formwork: 300 or more tonnes of fresh material which, thanks to the introduction of a special self-shaping metallic framework, takes on the final structural shape by means of the breath of air necessary to smoke a cigarette. But not only: this is not a simple experiment destined to remain in the mind of a dreamer, on a drawing board, or which at most has led to the construction of some *mockups*. The method developed by Bini back in 1964 has been used in more than twenty countries for the conception and subsequent realisation of an impressive number of until now still unknown buildings.[1] This is a construction method that is still today technically adopted to conceive and realise the thin cast shells designed in the United States by Nicolò Bini, Dante's son (Bini 2016).

Who is Dante Bini, the inventor of this particular way of building domes in thin reinforced concrete which, until now, represents the only way in the world thought up to search for a structural form and to construct it at the same time (Fig. 1.1)?[2]

[1] Bini referred to more than 1.600 shells constructed throughout the world through the system he invented; no precise survey has yet been carried out.

[2] The here-reported bibliographic and professional information about Dante Bini was taken from the volume written by Bini (2009), which was later revised and translated into English (2014), but also from a list entitled "Sequenza Temporale delle Attività" (Temporal Sequence of the Activities) drawn up by Bini himself and offered to the authors of this volume.

A. Pugnale and A. Bologna, *Architecture Beyond the Cupola*,
Mathematics and the Built Environment 7,
https://doi.org/10.1007/978-3-031-26735-2_1

Fig. 1.1 Dante Bini photographed at the "mushroom field", San Cesario sul Panaro, Italy, in February 2013 (WML)

Bini was born in Italy, in Castelfranco Emilia, on 22 April 1932. Coming from a family of entrepreneurs that worked in Emilia Romagna in the winemaking sector, he enrolled in the Faculty of Architecture at the University of Florence in 1952, where he graduated on 9 March 1962: his final dissertation was about a reconstruction project of an oenological plant in Marsala, which was characterised by a reinforced concrete dome to protect the bottling line.[3] During his studies, Bini worked in his family's business, where he also designed a packaging system for bottles that enabled him to win prestigious awards, such as the *Oscar for Packaging Design* conferred by the Istituto Italiano Imballaggio (Italian Packaging Institute) of Padua in 1962 and the *Eurostar for Packaging* in 1963 and 1964.

The year 1963 was a crucial year in Bini's career: with a friend he founded the "Old Home" society which dealt with the restructuring of old properties around Bologna and, at the same time, be began to work as a freelance architect, designing, with the

[3] The thesis was drawn up by Bini with Giuseppe Gori as Supervisor, Paolo Tincolini as Co-Supervisor and Pier Giovanni Garoglio and Ludovico Quaroni as External Supervisors.

collaboration of a local builder, various ex novo buildings. The construction which, according to Bini himself, he was most successful with is in Bologna at Number 8, Via dell'Osservanza: a four-storey, above-ground building characterised by cantilevered ceilings that determine stringcourse features and external walls covered in brickwork. The isolated residential townhouse is of a type of architecture that is widespread in the suburbs of Italian cities and is very different from that distinct way of "doing architecture" put in place by Bini following the intuition he had that would inexorably mark his professional career path: according to the legend, in the winter of 1963, Bini was playing a game of tennis in the sports facilities in the Giardini Margherita (Margherita Gardens) in Bologna: "a large green and white balloon, inflated at low pressure, had been set up to cover the playing courts: a so-called pressostatic (or super pressure) structure" wrote Bini (2014: 27). "At the end of an interminable match [...] we were no longer able to get out through the pneumatically sealed door of the 'balloon' since, in those few hours, Bologna had been covered by about a foot and a half of snow". As indicated in his writings, it was from this chance event that Bini had the construction idea that would later give shape to all the Binishell projects that would make him famous throughout the world: "as we shovelled away to clear a path through the snow, I wondered why I hadn't noticed increased pressure inside the 'balloon' caused by the heavy mass of snow that had gradually accumulated on the skin. I did a quick calculation of the approximate weight of the snow amassed on the surface of the structure which, inflated at low pressure, was supported by just a few hundredths of an atmosphere – in technical terms, only 3.45 kPa (0.5 psi) literally supported tons of weight!" (Bini 2014: 27–28). It was from this observation that Bini's great intuition arose: "I realised that the entire weight of the reinforced concrete dome that I had designed for my university dissertation could have been lifted with a pressure only slightly greater than that which was supporting the snow accumulated on that pressurized structure". In fact, apart from this account, as we will see later on in this chapter and also in Chap. 2, Bini's constructive idea did not represent an absolute novelty without any precedents that were already widely known at the time of his legendary game of tennis in 1963: indeed, it lies within the wake of the consolidated technologies through which, in the preceding twenty years, such designers of the calibre as Normand W. Mohr, Wallace Neff and James H. Marsh had already deposited patents and constructed numerous buildings in the United States, knowledge of which was widely published in the general press. However, Bini deserves to be merited with having courageously developed—in Italy, and in a completely independent and experimental way—the idea of generating architectural shapes and spaces through thin shells in reinforced concrete created, as we will see later on in detail, with a certainly more effective, pragmatic and widespread technique than that already exhibited by Mohr, Neff and Marsh, and which, for these reasons, has led to architectural results of greater importance.

Just a few months after the tennis game in the Giardini Margherita in Bologna, that is, on 4 July 1964, Bini erected his first experimental dome in Crespellano with an inflatable formwork. This was a building that was destined to house the Unipack headquarters, a company founded in 1961 by Bini for the designing and production of packaging for the bottles produced by his family's wine production company. Between

1964 and 1966, other domes were constructed between Pegola and the experimental site that Bini installed on some land he owned in San Cesario sul Panaro, near Castelfranco Emilia, a site that in those years was referred to as a "mushroom field", precisely because of the by-then widespread presence of domes and thin structures in concrete of various shapes and dimensions, which grew quickly, just like mushrooms.[4] In 1966, Bini founded—together with other investors from Milan, with a share of 30%—the *Binishells spa*, which is equivalent to a Public Limited Company (PLC). Convinced about the quality of his invention, and of its potential effect in entrepreneurial terms, Bini began to study and design, often in collaboration with other professional figures, dome constructions of different sizes and intended use.

Nonetheless, it was in fact the construction of the first dome in Crespellano that attracted the attention of a tutelary deity of thin vaults of those times: Mario Salvadori in fact by chance read the news of the erection of the dome, which had been published in a local newspaper from Modena that had arrived at the Italian Consulate in New York. A relationship of admiration and collaboration was set up between Bini and Salvadori that led to the construction of a demonstrative dome in the campus of Columbia University in New York in May 1967, which would bestow on Bini fame and visibility at a worldwide level (Whitehouse 1967).

"The years that followed that happy American experience saw me travelling the world promoting the Binishell system and presenting this new technology at conferences and lectures in locations from Eastern European [countries], to Central and South America. I also began working on the development and testing of new architectural forms using my construction system", wrote Bini (2014: 71). These were the years of feverish activity that witnessed the designing and construction of the famous villa in Costa Paradiso di Gallura, in Sardinia, which was built for Michelangelo Antonioni and Monica Vitti, conceived between 1969 and 1970 and completed in 1971: this is the building, which still today represents his most famous project, that Rem Koolhaas defined, on occasion of the XIV Biennale di Architettura di Venezia (the Venice Architecture Biennale), as "one of the best domestic architectures of the last one hundred years" (Docomomo 2021).

In the meantime, in 1970, following the break up with his partners from Milan, Bini was first ousted from his position as Vice President of *Binishells spa* and, in the same year, forced to leave the society. As a result of the contract with his ex-partners, Bini could autonomously use the construction system he invented only in Asia, Australia and the United States. It is also for this reason that, until now, it has not been easy or perhaps even completely possible to draw up a complete list of the Binishells created in those years. *Binishells spa* in fact constructed a substantial number of domes that is difficult to compute, not only in Italy, but also in Angola, Austria, Brazil, Cuba, France, Germany, Japan, Jamaica, Mexico, Iraq, Pakistan, Peru, Saudi Arabia, the UK and Venezuela. A list of these works and of all the boundary conditions through which these architectures were conceived and constructed cannot but be the result of

[4] According to Bini, the term "mushroom field" was coined by Salvadori during his visit to the early Italian Binishell prototypes. In Italian, the mushroom field is referred to as "la fungaia". The full address of the mushroom field is: Via Castel Leone, San Cesario sul Panaro, Modena, Italy.

a complex and rigorous research which, it is hoped, can be conducted in the future, with the certainty that this will highlight further potentialities and complexities of the Binishell system according to the different local conditions of its application.

Furthermore, it is difficult to attribute an exclusive Bini authorship to each of the Binishells designed by *Binishells spa* before 1970 because of the multitude of applications of the construction system made in collaboration with other colleagues, both engineers and architects, although the authorship of his patent has never been in doubt. As we will see in the next chapters, among the professional figures who were brought in by Bini and by *Binishells spa* for the designing of domes of various types, the following can surely be mentioned: Michael Godwin and John Faber, Alexandre Jeleff, Furio Nordio, Vittorio Casini, Riccardo Merlo, Mario De Franchis, Anna d'Alessandris Pazzi with Angelo Berardi, the De Franchis-Verni firm in Florence, the Edilizia Mediterranea firm in Naples, the Bonfiglioli, Evangelisti and Vacchi firm in Bologna, the Ufficio Tecnico Concari in Parma and many others.[5]

Bini continued to carry out the role of external consultant for *Binishells spa*; in the meantime, in 1970, he founded, together with the architect Eros Parmeggiani, the "Dante Bini & Associati" firm of architects, with whom he would design various villas in Sardinia, in the Costa Paradiso and in the Emerald Coast until 1972, projects in which he took charge of the construction site works.

For Bini, the 1970s represent the decade in which he was able to make a significant breakthrough in his career, thanks to the application of his invention in Australia. He reported how "at the beginning of autumn 1971, the Agent General of New South Wales, Davis Hughes in his London office, received an unusual request from the minister of Public Works, back in Australia, the Honourable Leon Ashton Punch. The request was to ascertain the quickest building system being used in the UK to construct multi-purpose centres, gymnasia, libraries and other school buildings that New South Wales desperately needed and that the very same minister had promised during the electoral campaign. The new facilities that were promised had never been built, with consequent political embarrassment" (Bini 2014: 74). The Binishell turned out to be the most suitable construction system for a rapid realisation of these buildings: after the visit of a Ministry delegate to Italy and a preliminary journey of Bini to Sydney in 1973, the architect, accompanied by his wife Adria and his two sons, Stefano and Nicolò—the latter, as we shall see throughout the entire volume, is today working in the United States where he develops his building projects in thin reinforced concrete shells—then moved to Australia (Pugnale and Bologna 2015: 360–362). There, he further developed his construction techniques and his architectural mannerisms, and he advanced the technical progress of the pneumatic lifting systems of cast-in-place concrete structures, constructing more than twenty buildings with 18- to 36-m diameter domes. Among the most well-known constructions in Australia, designed by Bini himself or under his direct supervision, it is worth mentioning the Narrabeen North Public School Binishells, examined in detail in

---

[5] Various *Binishells spa* promotional catalogues are kept in Dante Bini's private archive in St. Helena, California, USA, and were made available to the authors of this volume for the conduction of their research.

Chaps. 4 and 5, the Georges River College multipurpose centre in Peakhurst and the Space City Shopping Centre in Kallangur. As Bini was not enrolled in any of the Australian Professional associations, in 1974, Bini founded "Bini Consultants Australia" together with the Taylor Thomson Whitting (TTW) engineering consultancy firm, with the intention of taking control of the Binishell system throughout the whole of Australia; in April 1976, he signed an agreement with the *Jennings Industries Ltd* in Melbourne and thus was able to spread his construction technology throughout the entire continent.

From the end of the 1970s, after moving permanently to the United States, he extended his research to other forms of construction automation by applying the "self-shaping" principle of architecture[6]—in other words, constructions obtained through resort to pneumatic formworks—to different types of structures, ranging from geodetic ones in steel to prefabricated ones in wood, not to mention the invention of other pneumatic structures, which were even thought up to be used on the moon.

Bini is still working today, at 91 years of age, on the perfectioning and constant actualisation of the construction techniques necessary for the realisation of his Binishells. In the year of writing this volume, he is in fact still at the thick of designing, experimenting and depositing a patent for a new type of reinforcement: his ambition is to face the enormous challenge of constructing prestressed reinforced concrete thin shell structures, obtained through the use of reinforcements made up of stainless steel springs or in carbon fibre that are able, again thanks to the use of pneumatic formworks, to put the whole concrete structure under compression, thereby completely eliminating the iron rods of the reinforcement.

In the light of these premises, this introductory chapter has the aim of investigating the relationship between the constructive technology and the architectural and structural form of the buildings created by means of the Binishell system. The evolution of the construction process patented by Bini is presented and, on the basis of the technical requirements, the architectural research that has been derived from it, in terms of composition and form, is also illustrated. Our intention is to introduce the Binishell in the frame of a design culture that is based on construction, through the relationship that exists between tectonics, compositional aspects and conceptions of the architectural space.

1.2 At the Roots of the Binishell System

It is likely that the idea of creating monolithic structures in concrete by means of pneumatic formworks can be attributed to the architect Normand W. Mohr who, in February 1927, proposed realising the Golden Gate Bridge in San Francisco as an

[6] This is an Italian term that Bini uses in a recurring way to specifically indicate buildings constructed by means of construction methods he invented to resort to the use of pneumatic formworks and of reinforcements that are capable of progressively taking on the shape of a formwork as the latter is inflated. *L'Architettura Autoformante* is also the subtitle of one of his books published in 2009.

underwater tunnel of reinforced concrete. It should have been built using a pneumatic formwork and that probably appeared extraordinary to the readers of the American *The Architect and the Engineer* magazine.[7] Mohr was experienced in the manufacturing of aerostatic balloons: in April 1916, he filed a patent for a *Combined aeroplane and dirigible balloon,* in which the characteristics, potentialities and construction procedures of inflatable structures were clearly mastered (Mohr 1916). Despite his intuition of applying such a system for the erection of circular buildings, his ideas fell into oblivion for about six years.[8] In August 1933, *The Architect and the Engineer* was still giving space to Mohr's invention by publishing a short article in which, through a sketch, he further detailed his construction idea.[9] The steel reinforcement, as conceived by Mohr, was of the traditional type and assembled above the pneumatic formwork just after it was inflated. Only in 1964, was Bini able to introduce a substantial innovation, that is, of creating his concrete shells with an extensible steel reinforcement, which was mounted above a deflated pneumatic formwork, and then deformed to take shape following inflation of the membrane.

The assumptions made by Mohr were successfully developed again by the Californian architect Wallace Neff, whose reinforced concrete shell houses, built with inflatable and reusable pneumatic formworks, were a great success (Neff and Clark 1986: 177–186). In 1934, Neff made the first drawings of a pneumatic house and, in 1939, he wrote a description of his pneumatic form (Head 2011: 18). Neff's patent, which was filed in April 1941, represents the result of the socio-political environment of that period, which also led other architects to work on the same topic and develop low-cost houses or "refuges". The aim behind the patent was to build thin shell concrete barrel- and dome-type structures for dwellings, hangars and barracks for defence housing purposes. Neff planned to insert a series of hooks into a concrete footing slab, to which he would anchor an inflatable membrane that could be shaped to assume a desired size and curvature (Fig. 1.2).

Neff's form was fabricated from an airtight and substantially non-stretchable material of sufficient strength to withstand hard usage, such as a rubber-impregnated canvas. Once inflated, the formwork was covered with a wire reinforcing mesh of an approved type; thus, with the so-called gunite method, a continuous layer of about three inches in thickness (a mixture of cement, fine aggregates and water) was literally shot against the inflated membrane through a cement gun, in such a manner that the wet concrete retained its original position. The concrete passed across the

---

7 By the latest system of pneumatic, inflatable and deflatable forms, dock pouring, launching, floating to position and submerging in place, submarine tubes may be built at one-third the cost of suspension bridges (Mohr 1927: 83).

8 Curved in sky line in mass and detail, facilitated by improvement in method of construction (notably the pneumatic form [...]). By this method buildings or circular form will be more economical to construct than our present rectangular system. Also being more closely natural in form, more beautiful (Mohr 1927: 84).

9 After plaster shell is set, the rubber form is deflated and is removed with canvas-casing, to be used over and over again. The insulation paper which prevents plaster from adhering to canvas-casing is washed away, leaving a finished blaster interior (Mohr 1933: 4).

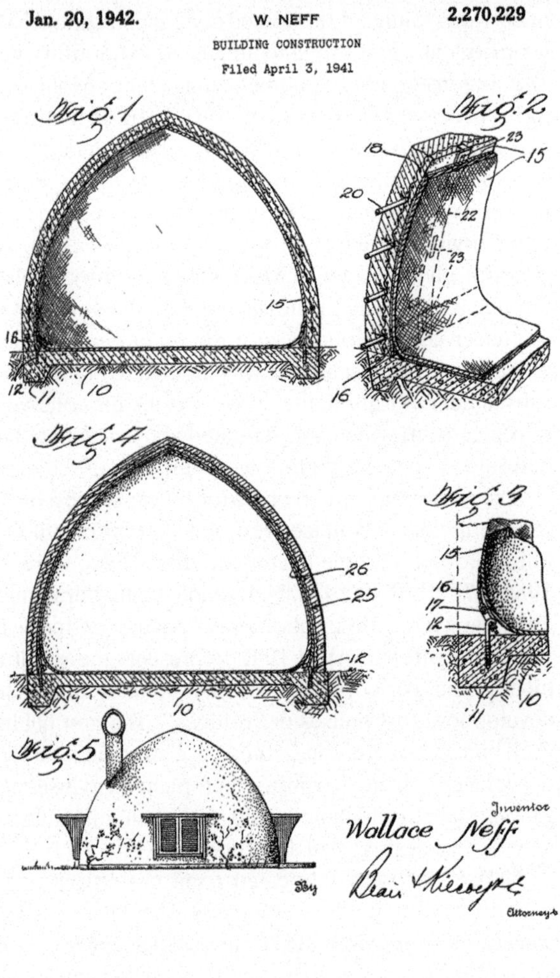

Fig. 1.2 Wallace Neff, *Building construction*, patent US 2,270,229, filed on 3 April 1941. Drawings

iron mesh and flowed directly against the rubber surface of the formwork. Prior to inurement, the exterior face of the shell was levelled with rods to obtain the final curvature and a smoother surface. After hardening the structure, the membrane was deflated and then removed through one of the openings that had been created. The subsequent patents filed by Neff between November 1941 and June 1952 (Fig. 1.3) were aimed at improving the concrete casting, at the manufacturing of an economic and reusable pneumatic formwork (Neff suggested nylon, cotton muslins or treated paper of the type developed at that time for parachutes), and at the positioning of steel reinforcements (Neff 1941a, b, 1942, 1952). Through the employment of such a construction system, by 1941 Neff had built twelve experimental houses in the so-called *mushroom village* in Falls Church, Virginia (*Life* and *The Architectural Forum*, 1941). In 1945, once he had understood the potential of the invention, Neff founded a construction company, called *The Airform International Construction Company,* through which he built hundreds of reinforced concrete shell structures (Fig. 1.4) all over the world (*The Architectural Forum* 1947 and Muntz 1989).

During the Fifties, there were no further innovations in this field. The shell structures that arose from the collaboration between Neff and the architect Eliot Noyes were erected using a by-then well-established construction technique which, in the United States, allowed low-cost houses to be produced quickly (*Life* and *Progressive Architecture,* 1954). In 1954, Salvadori worked with Neff and Noyes to improve the casting of in situ concrete through the use of a "concrete gun" (Scipio 1967: 60).

However, it should be underlined that the popularity of these formalisms induced a certain fascination in this newly established bubble architecture by both designers and by the customers and the operators in the construction industry in America at that time: this is why, starting from the 1950s, the idea of being able to transform an inflated ball into a place to live began to take shape. The Fiberthin Village project, elaborated by Frank Lloyd Wright in 1956 for the U.S. Rubber Company

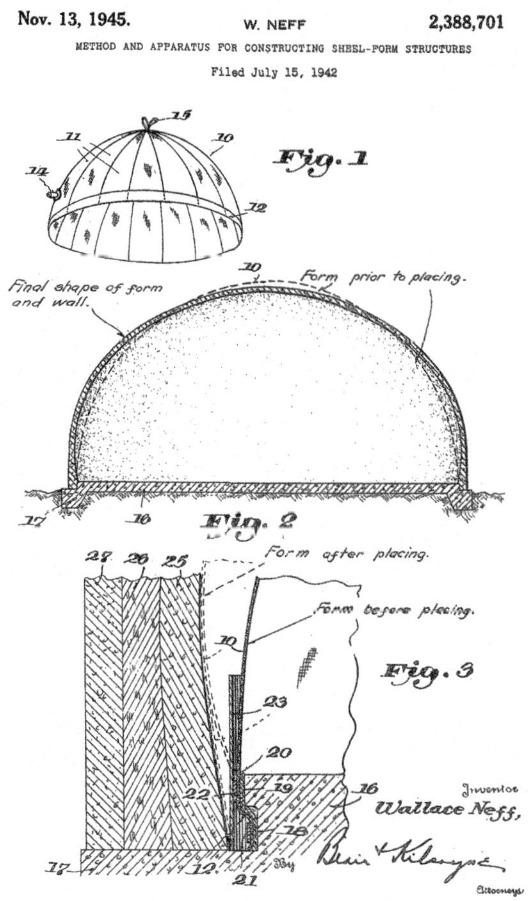

Fig. 1.3 Wallace Neff, *Method and apparatus for constructing shell-form structures*, patent US 2,388,701, filed on 15 July 1942. Drawings

Fig. 1.4 Wallace Neff, an "Airform" concrete shell house, November 1941. Building site (MLP, HLSM)

and thought up to be built in Mishawaka, Indiana, in fact falls into this design culture (Fig. 1.5); designed in the same year as the huge concrete domed shell of the Annunciation Greek Orthodox Church for the Milwaukee Hellenic Community (Pfeiffer and Goessel 2015: 462–465), the Fiberthin Village featured a series of 7.62-by-14.02-m hemispherical affordable homes for the common person, using the Fiberthin system, a vinyl-coated nylon fabric. This was a way of "doing architecture" that remained a single event in Wright's production, but which was about 20 years ahead of the much more renowned project of José Miguel de Prada Poole for the Instant City of Ibiza in 1971, the Pamplona Encounters in 1972 and the Instanlo Hielotrón ice skating centre in Seville in 1975.

Looking again at the structures in reinforced concrete, the Swiss engineer Heinz Isler, in parallel with Neff and Noyes' constructions in the US, exploited the proprieties of pneumatic structures in Europe starting from the summer of 1955. He used scale models as form-finding tools to derive the structurally optimal geometries in order to build shells with a minimum of material (Chilton 2000: 34–37). However, the innovation of Isler's design process did not simplify construction, and it was still necessary to use complex timber formworks to build his concrete shells (Boller and Beckh 2019; Bini 1984: 189).

Thus, we arrive at the 1960s, the decade that witnessed the introduction of the Binishell system: the first experimental dome was built in Crespellano, near Bologna,

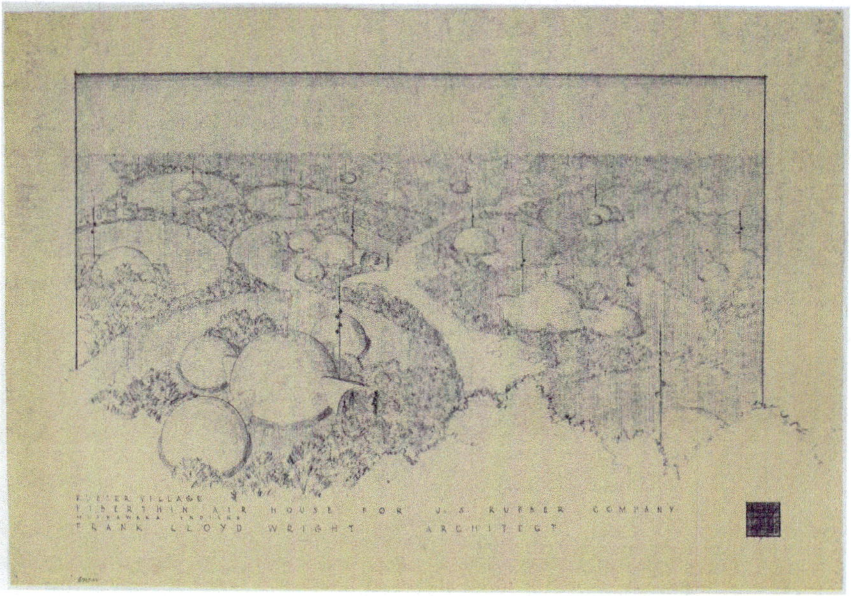

Fig. 1.5 Frank Lloyd Wright, Fiberthin Village project (unbuilt), Mishawaka, Indiana, 1956. Perspective view (FLWFA)

in July 1964 (Fig. 1.6).[10] This was the first large-scale application of the Binishell patent, which was then perfected over the years (Figs. 1.7, 1.8 and 1.9).

The Binishell system requires the preparation of a footing ring beam. A membrane is then placed on top of the formwork. According to Bini, a reinforcement system, consisting of metal springs and bars, is arranged by following a geometry that was empirically inspired by the planimetric projection of the cast-in-place ribs of Pier Luigi Nervi's dome in the Palazzetto dello Sport in Rome (Fig. 1.10).[11] The springs have the same function as a bar chair, i.e. they keep reinforcement bars at the correct height within the shell, and are particularly useful during the inflation process, which bends the bars into their final form. Bini's framework was therefore able to deform from the ground to a tri-dimensional shape of which the planar configuration was

[10] Bini systematised the design process of his inventions in his books published in 2009 and 2014.

[11]
> It was important that the reinforcement rods, used to resist tension (unlike concrete which only resists compression) once lifted, would attain the position determined by my engineering calculations derived from the study of Nervi's elegant sports arena, the Palazzetto dello Sport in Rome. Applying some creativity, the concrete metal reinforcement system of the first dome was composed of coaxial circles of steel rods, not actually engineered by design, arranged on the deflated membrane while it was still on the ground. The distance between the circular rods was secured by chain segments, which would obviously not pose any resistance to their shorter distribution on the ground. When lifted from the ring beam base, these chains would arch up in broad curves towards the pole in opposite directions, following the exact geometry of the ribbing of the Palazzetto dello Sport (Bini 2014: 28–29).

Fig. 1.6 Dante Bini, Crespellano Binishell, Unipack company, July 1964 (DB)

just a projection; the anchoring system of the inflatable membrane was comparable with that used for fixing traditional pneumatic envelopes (Herzog 1976: 148–154).

A reinforced concrete dome—or, as Bini stated, a shell structure in ferrocement—was obtained after sixty minutes of membrane inflation.[12] During erection of the formwork, concrete was restrained by a net fence. For this reason, it appeared grainy and not compacted at the final stage. An acceptable smooth surface was obtained by means of hand levelling at the extrados with a trowel, without applying any mechanical vibration. The introduction of iron springs as a reinforcement in the framework (the rods were inserted inside the springs and were only anchored to the ground at one end, and were therefore free to take on the shape given by the dome) took place in the fourth experimental dome, which was built by Bini in Pegola (Fig. 1.11). Between 1964 and 1965, Bini conducted new experiments and with his company *Binishells spa*. A 30-m diameter dome was built at the "mushroom field" (Fig. 1.12), where, with the use of a lightweight reinforced concrete, he obtained "Leca-type" aggregates (Bini 2009: 40) and a new type of framework, characterised by a spiral shaped mesh, and interlaced with structural steel rods and bars (*Binishells spa* 1965; Bini 1966a, b).

[12] In his writings, Bini places his invention in the wake of the domes built by the most renowned designers of the last Century: the use of the Italian term *ferrocemento* (ferrocement) to indicate the material his shells are built with may still indicate his debt to the work of Nervi, who, in 1943, patented a material called *ferrocemento* with which he realised his well-known prefabricated domes (Bini 2008: 124).

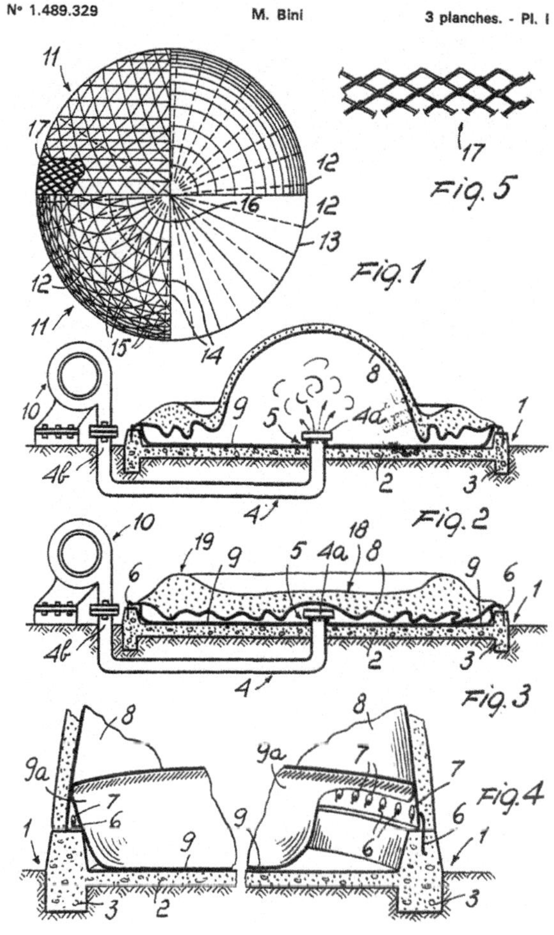

Fig. 1.7 Dante Bini, *Méthode pour l'érection de bâtiments constitués principalement d'une structure portante en une seule pièce, équipement pour réaliser cette méthode, et bâtiment ainsi obtenu*, patent FR 1489329, filed on 9 August 1966. Drawings of the first Binishell pneumatic construction system

When constructing Binishells, concrete is poured onto the ground, and the reinforcement system placed above a deflated pneumatic formwork is submerged. Once the concrete has been poured, air is pumped into the edge-beam/membrane system, and a dome forms within 1–4 h. Concrete is compacted, using surface vibrators that act on the visible side of the dome, and it is then levelled by hand. The drawing attached to Bini's patent filed in Italy in November 1968 illustrates that only three workers were required to complete this process (Bini 1968).Following Salvadori's advice, probably in early 1967, a second PVC membrane was added to the system to cover the concrete while it was still wet and flat on the ground (Fig. 1.13). The introduction of this double membrane facilitated compaction and resulted in a smoother outer surface with fewer imperfections (Bini 1966a, b; Bailey 1967; *The Architectural Forum* 1967); opening cuts were only made once the concrete had hardened and the pneumatic formworks had been removed and stored for reuse.

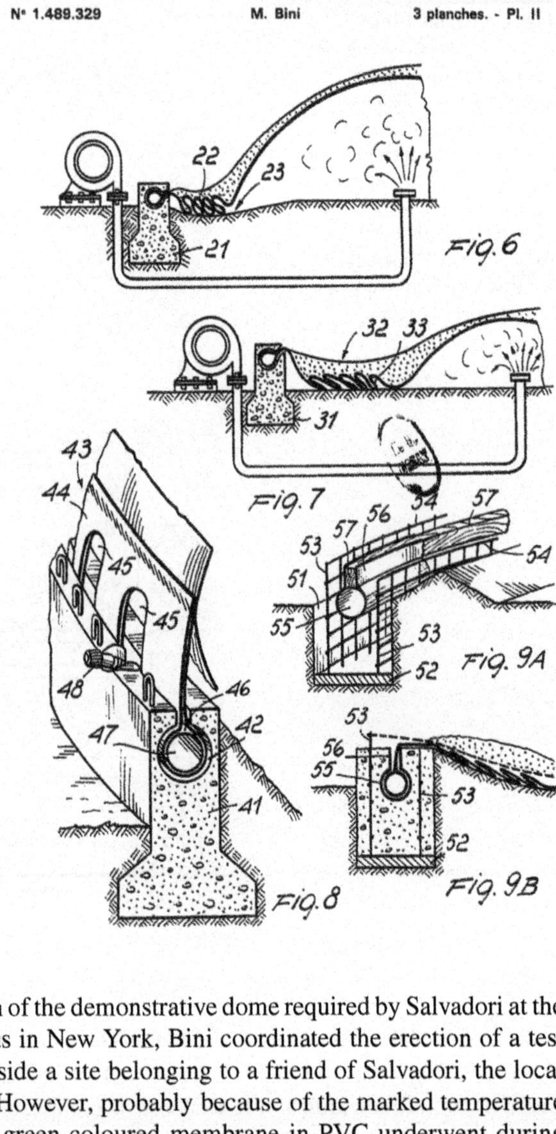

Fig. 1.8 Dante Bini, *Méthode pour l'érection de bâtiments constitués principalement d'une structure portante en une seule pièce, équipement pour réaliser cette méthode, et bâtiment ainsi obtenu*, patent FR 1489329, filed on 9 August 1966. Schemes and drawings of the concrete edge ring

In view of the construction of the demonstrative dome required by Salvadori at the Columbia University Campus in New York, Bini coordinated the erection of a test dome in Reston, Virginia, inside a site belonging to a friend of Salvadori, the local builder Robert E. Simon Jr. However, probably because of the marked temperature fluctuations that the second green-coloured membrane in PVC underwent during its transport by plane from Italy, it tore when being lifted (Figs. 1.14 and 1.15). Therefore, a new, thicker and grey-coloured membrane was retrieved in America for use during the event organised by Salvadori in New York, which was planned for between 16 and 17 May 1967 (Fig. 1.16). Some days before, three men from the Rissil Construction Company constructed the foundation ring in the site in front of the Pupin Physics Laboratories, and the American grey membrane seemed to have a too high tensile strength, recalled Bini. "We decided to risk rolling up the two membranes together, in such a way that, when the plastic tube that contained both of

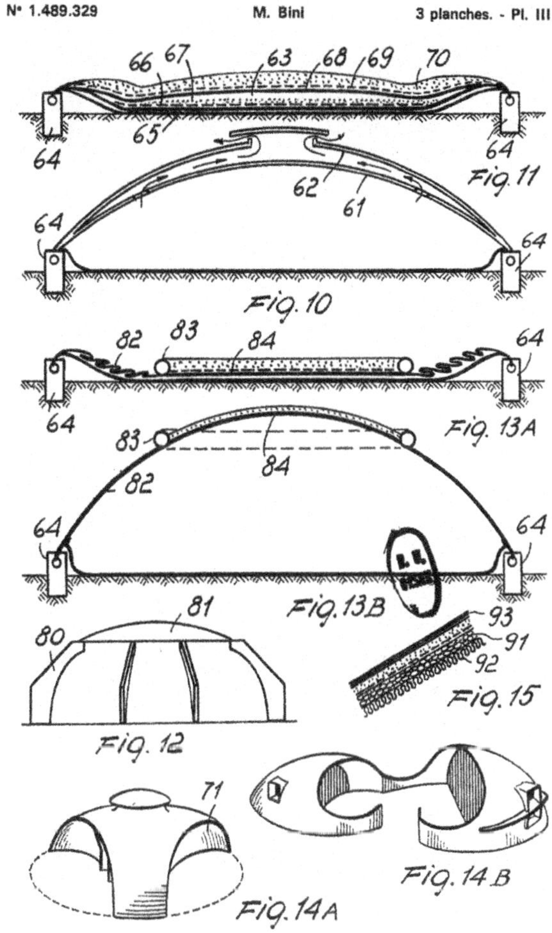

Fig. 1.9 Dante Bini, *Méthode pour l'érection de bâtiments constitués principalement d'une structure portante en une seule pièce, équipement pour réaliser cette méthode, et bâtiment ainsi obtenu*, patent FR 1489329, filed on 9 August 1966. Schemes of the pneumatic construction system including potential design solutions

them had been unrolled on the just poured cement, the green membrane would have remained outside and the grey one inside in contact with the concrete" (2014: 62). But even in this case, the green membrane that arrived from Italy was damaged. Thanks to the shrewdness of Bini and Salvadori, and the alertness of the trusted workmen that Bini had brought from Italy for the occasion, almost none of the people present were aware of the problem: "[…] the outer membrane began slowly and inexorably to open from the pole, along the line of adhesion. With the speed of lighting, my three assistants cut the membrane at the base", as per Bini's suggestion "and removed it immediately" (2014: 63).

During the *1st International Colloquium on Pneumatic Structures*, which was organised at the University of Stuttgart, Germany, on 11–12 May 1967, by the International Association for Shell Structures (IASS), and which took place just a few days before the construction of Bini's dome in New York, Isler stated that "it would

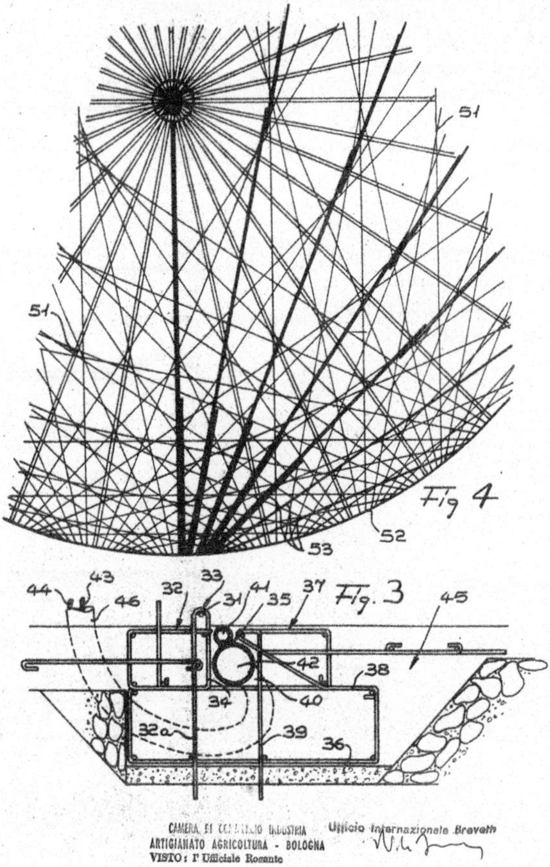

Fig. 1.10 Dante Bini, *Metodo per la erezione di edifici e strutture a membrana monolitica perfezionato nel complesso della armatura estensibile nonché nella compattazione del materiale*, patent IT 853736, filed on 21 November 1968. Drawings of the steel reinforcement

be logical to use pneumatic membranes directly as formwork" (1967: 51). However, he also noted that, at the time, there were still great difficulties involved in doing that. Bini challenged Isler's statements in the same colloquium and presented his Italian shell structures. The publication of the proceedings included his contribution with photographs of the latest achievements on the use of pneumatic formworks, already completed with the second outer PVC membrane (Bini 1967).

1.3 Constructive Pragmatism and Empirical Experimentation

Bini's technical inventions led to the construction of buildings that unequivocally belong to a particular season of the design culture of the twenty century. It is therefore

Fig. 1.11 Dante Bini, Pegola Binishell, 1964–1965. The building under construction (DB)

Fig. 1.12 Dante Bini, 30-m diameter dome at the “mushroom field”, San Cesario sul Panaro, 1964–1965. Photograph with Bini’s and Salvadori’s cars (DB)

crucial to analyse such inventions—and the resulting structures—from an architectural perspective, thus going beyond a mere discussion on construction means and methods.

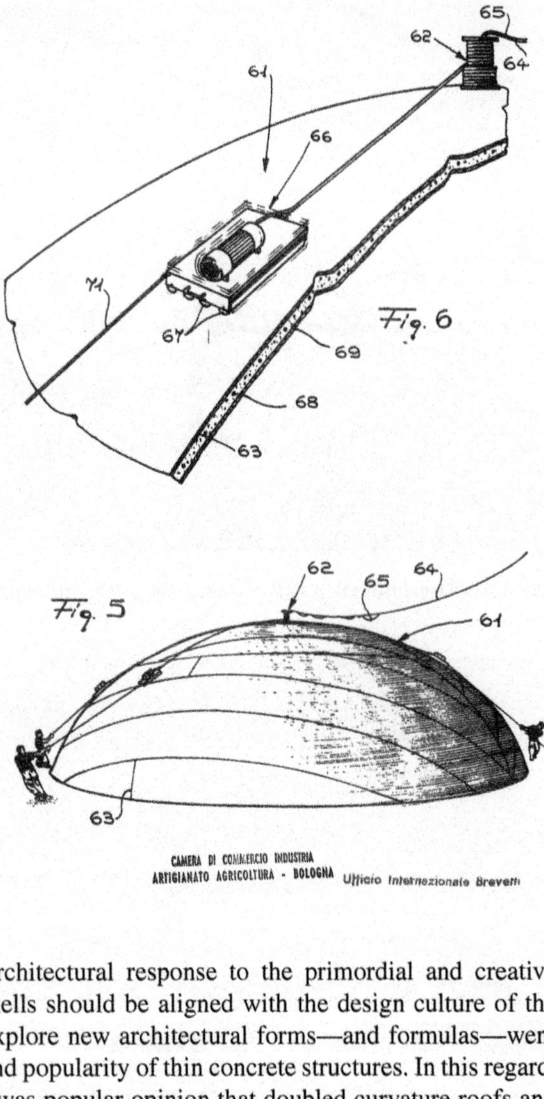

Fig. 1.13 Dante Bini, *Metodo per la erezione di edifici e strutture a membrana monolitica perfezionato nel complesso della armatura estensibile nonché nella compattazione del materiale*, patent IT 853,736, filed on 21 November 1968. Schemes of the concrete compaction technique

According to Bini, his architectural response to the primordial and creative impulse to shape concrete shells should be aligned with the design culture of the 1960s, where the desire to explore new architectural forms—and formulas—went hand in hand with the trend and popularity of thin concrete structures. In this regard, Bini wrote: "in the 1960s, it was popular opinion that doubled curvature roofs and thin shell structures represented a new building standard", which was seen as an occasion to design and combine "the most innovative scientific formulae related to architecture and construction sciences" (2014: 26).

This ascending and descending interest in thin shell structures should be taken into account, as it initially provided a fertile environment for the development of Binishells, in the 1960s, and then progressively decreed their obsolescence. Therefore, it is easy to associate Bini's Binishells with iconic images of the shell and spatial structure designs of Heinz Isler, Félix Candela, Frei Otto and Richard Buckminster

Fig. 1.14 Dante Bini, Binishell construction demonstration, Reston, Virginia, April 1967. Top: concrete pouring phase; Bottom: workers placing the outer membrane over the fresh concrete (DB)

Fuller. Despite the obvious link with international design precedents, Bini's structures can be considered the direct result—and probably the last tangible example—of designs of concrete structures obtained from empirical experimentation. This is an approach that originated in Italy with Giovanni Antonio Porcheddu and Attilio Muggia pioneers responsible for the introduction of the Hennebique system into Italy—and was then exported worldwide by Nervi.

It was not by chance that Bini cultivated a hybrid profile of designer, builder and entrepreneur in the 1960s, during the Italian economic miracle. Bini developed his construction techniques in-house by following a very pragmatic approach. His work can be considered the result of experience acquired through professional practice and experimentation carried out directly at the construction site. Bini's buildings are neither the product of a method that privileges structural analysis nor design concepts based on the reinterpretation of architectural types and design precedents.

Site photographs and archival sources that describe structural tests and analyses conducted on the first experimental Binishells—those built at Crespellano, Pegola and Castelfranco Emilia—reveal the high degree of empiricism that characterised Bini's design approach (Figs. 1.17 and 1.18). His was a way of finding structural form

Fig. 1.15 Dante Bini, Binishell construction demonstration, Reston, Virginia, April 1967. The torn outer membrane at the end of the inflation process (DB)

from an Italian culture that was based on practical building experience and prototyping. Such construction experience was essential to achieve a degree of technical innovation, which in turn informed the development of new architectural forms. At the time, it would have been difficult to attain similar results in Italy using numerical structural analysis tools: form-finding was still performed using experimental methods, i.e. scale models (Bini 2014: 37–38).

Within this framework, it is no coincidence that Binishells were developed by an architect rather than by an engineer: "if you were an engineer you would never have conceived such nonsense" (Bini 2014: 59) was Salvadori's comment on Bini's invention. It is a revolution in the redefinition of the relationship between architectural form and construction system: neither of the two parts of the equation takes priority over the other. This approach to designing inflatable concrete shells is still visible today in the work of Bini's son, Nicolò Bini, who is now the CEO/president of the US *Binishells Inc.* company, where he continues to experiment with pneumatic construction methods for concrete shell structures and to promote several techniques based on pneumatic membranes to erect concrete shells rapidly while reducing waste: this is the case, for instance, in his design of a house in Malibu for Robert Downey Jr, which is discussed in Chap. 6.

Fig. 1.16 Dante Bini, Binishell construction demonstration, Columbia University, New York, 16 May 1967. Photograph of the inflation process (DB)

Bini's first Italian patent on the construction of reinforced concrete domes using a pneumatic formwork dates back to 1964. A following patent, which Bini filed in the United States on 12 December 1966, describes the Binishell system as a: "Method for erecting domelike and other structures", which employs a sheetlike inflatable membrane to achieve a desired structural shape. The patent includes the steps necessary to build the shell through a pneumatic formwork and describes how to position the reinforcing bars and pour and compact the concrete (Bini 1966b).

It is clear from the patent drawings and descriptions that using an inflatable construction technique affected Bini's design process of determining "new architectural formulae". Technical and technological innovation determined a structural form, which, consequently, informed the architectural form. But perhaps it was the other way around: the ambition to design large domes and Bini's attitude towards architectural design influenced him, so that he selected a specific structural typology first and, as a consequence, this triggered the development of a new form-finding and construction technique. The two aspects are so closely related to each other in Bini's work that it becomes impossible to determine a clear hierarchy.

Bini's new construction system is a prefabrication technique since it reduces the construction time and the workforce required to build a finished shell. Many other precast concrete systems were used in the Italian construction industry between the 1950s and 1960s, such as Nervi's ferrocement system. However, Bini succeeded in improving most aspects of those systems, thanks to his ability to condense, in a single invention, which he called the "three great expressions of construction science" that

Fig. 1.17 Dante Bini, early Binishell prototypes at the "mushroom field", San Cesario sul Panaro, March 1968. Construction sequence (DB)

"catalysed those years" and that "simultaneously dominated the range of architecture and advanced engineering that everyone studied at university". According to Bini, this is a combination of "thin shell structures, geodesic domes and tensile structures" (2014: 26): regardless of the technical appropriateness, in absolute terms, of this statement with reference to the conceptual, structural and material nature of his invention, it is certain that the Binishell illustrates the perfect union of architectural, structural form and construction technique. It defines a design culture in which, essentially, those two design aspects coincide.

Fig. 1.18 Dante Bini, an early Binishell prototype at the "mushroom field", San Cesario sul Panaro, March 1968. Shell surface smoothing (DB)

1.4 Bini's Architectural Formalisms

The architectural genesis, language and spatial planning of Bini's early domes depend on—and are closely linked to—the invention of the Binishell system. The limits of such a system surely imposed rigid construction constraints and the use of a specific typology, but, on the other hand, the Binishell system allowed Bini and other architects who used his system to experiment with a variety of spatial layouts and shell opening shapes.

Some apparent similarities with serial production methods are at the base of Bini's invention, and it is such a characteristic that made Binishells so successful in both Italy and abroad. The system was exported to all the continents, as demonstrated by some of the promotional catalogues and by the variety of patents filed by Bini throughout the world. The demonstrative prototype built at Columbia University for Salvadori is worth mentioning, since it led Bini to achieve international recognition. The Australian Binishells have also played an important role, since Bini decided to move to Sydney to build various 36-m domes. These domes had a height of 11 m at the apex and were to be used as school libraries and sports centres. Bini pointed out that the serial reuse of the same formwork was important for the Australian Binishells, as well as the limited number of workers required to prepare, lay, dismantle, handle and clean the two membranes.[13]

As discussed in the subsequent chapters, from the architectural point of view, Bini focused on four aspects when designing his early dome shells: (1) the size of the dome, (2) the connection to the ground, (3) the position and shape of the openings made on the concrete shell to house doors and windows and (4) the possibility of intersecting a Binishell with other identical or similar domes. Once he had validated the new construction method, spatial planning came into play. It is worth highlighting that a geometrical imposition constrained the challenge he faced of designing spaces within a circular building. The form was not informed by the programmes of the building: it was the typology and layout that were defined from the shell geometry.

Bini's invention is architecturally simple but technically honest, as the construction process of a completed Binishell can be observed from both inside and outside. As highlighted in Chap. 2, the rough inner surface of a Binishell is marked by the shape of the reinforcement, which is pushed to the edge by the pressure of the inflatable membrane, thus giving it the connotation of being a new archetypal hut. The exterior surface also reveals its own nature because of the way the openings are cut. Binishell is an invention that leads the designer to generate space from the relationship between inside and outside. In most cases, Bini focused on the quality of the interior, which in turn affected the architectural form visible from the outside.

A single shell does not offer a broad range of compositional opportunities to an architect, so Bini soon began to experiment with the combination of Binishells. As indicated in Chaps. 3 and 5, a potentially infinite number of layouts can be created by composing and intersecting Binishells of different diameters and heights. One of the *Binishells spa* promotional catalogues distributed in Italy states that "by arranging the metal reinforcement in various ways and modifying the constraints at the base, infinite variations in shape can be obtained, creating decorative elements and static reinforcement at the same time". In this way, "there are practically no limits in plastically sculpting the spaces defined by the Binishells, inserting and removing volumes and surfaces, in a succession of sensations that stimulate the inventive sensitivity of the designer".[14] Accordingly, Binishells can provide an architectural response to utopian spatial dreams, such as that expressed by Frederick J. Kiesler in his Endless House or by John M. Johansen in his Spray Concrete House #2, both discussed later on in Chap. 6.

The serial use of the same pneumatic formwork has proved, over time, to be the main limit of Bini's invention, if considered from the point of view of freeing architectural form from the constructive constraints inherent to monolithic reinforced concrete construction. The assembly of concrete shells of different sizes and the

[13] Bini's criticism of the system pertains to the difficulties that the twenty workers encountered in the manoeuvring, positioning, anchoring and accurate folding on the membrane formwork. He wrote: "in addition to being very expensive, [the membrane] had a surface area of 1000 square metres (10,700 square feet), weighed over 1,500 kg (1.6 tons), and had to be meticulously clean" (Bini 2014: 101).

[14] The *Binishells spa* promotional catalogue kept in Bini's private archive in St. Helena, California, USA.

creation of structures by means of their intersection will surely have been difficult operation and certainly not an economic one. Therefore, Bini's patent applications have typically been limited to constructing buildings with single domes for different uses, whose functional layout was developed within the spatial limits defined by the circumference of the perimeter of the dome itself.

From a purely typological point of view, Bini's result was not so different from what Neff and Noyes had achieved a few years earlier in the United States with their "bubble houses". The strength of Bini's invention is that the construction process can be standardised, which means reusing the same pneumatic formwork several times to erect domes that share the same dimensions. It is a standardisation that also allows conventional casting procedures to be used, since a Binishell is poured onto the ground as a flat slab. This sort of standardisation also invites a designer to experiment with the repetition and geometric intersection of Binishells, so that buildings with larger footprints than the dome diameter, such as sports centres, can still be covered using this system.

It is no coincidence that one of the *Binishells spa* promotional catalogues, printed between 1969 and 1970, advertised Bini's construction system—for the Italian market—as the answer "to the need for building structures of extraordinary versatility, flexibility and economy" in the construction of buildings "serving industry and agriculture". In the construction world, "there is an increasing need to reduce the cost of buildings, which generally exceeds by many times the cost of the plant and equipment for which they are built". Therefore, the most economical solution is the construction of single-dome buildings, "free of constraining infrastructures and load-bearing elements".[15] The numerous *Binishells spa* promotional catalogues, which showed photographs and drawings of a series of buildings already constructed at that time, and possible architectural and structural solutions that could be obtained through the application of Bini's invention by other designers illustrate the almost exclusive use of single shells, albeit for different uses. As will be seen in Chap. 5, Bini's experimentations, which were intended to improve his original invention, led him to develop a number of construction systems based on the use of air. Among these systems, it is worth mentioning a rather unique modular prototype that can be used to build industrial sheds with a unit roof on a square plan: a single pneumatic formwork, inflated on site, capable of producing an unlimited number of curved slabs, which, once hardened, could easily be placed on top of a square structural grid of columns, also presumably prefabricated (Fig. 1.19). To the best of our knowledge, this prefabricated industrial roofing system has never been studied in any depth and has not been well documented: a diagram in a *Binishells spa* catalogue and a photograph of a built prototype are all that are now available. This is a technical solution that seems to disregard the design factors related to obtaining spatial quality, and it was probably developed to maximise speed and economy of construction in the production and assembly of prefabricated concrete structures for industrial use. The rationality and simplicity of this modular system on a grid of columns can be compared with that developed by Nervi and Aldo Arcangeli for the construction of the floor formworks

[15] Quotation taken from a *Binishells spa* promotional catalogue (DB).

of the Gatti Wool Factory in Rome (Gargiani and Bologna 2016:217–243). Despite being structurally and formally different, the two systems share a similar pragmatic approach to the design of a unit roof element on a square plan supported at the four corners. However, it is not easy to make a detailed comparison between the two, particularly when commenting on the architectural space generated by Bini's invention, since this construction technique did not evolve beyond the prototype stage: Bini constructed a single span unit in his "mushroom field" near Castelfranco Emilia.

Bini's designs based on a series of domes were made by replicating individually inflated shells, placed side by side, i.e. without their volumes intersecting. This is also the case of an industrial building in Imperia, Italy, designed by the architect Furio Nordio (based in Trieste), which consists of two 30-m diameter connected Binishells placed side by side (Fig. 1.20). However, the aforementioned catalogue mainly presented a series of projects rather than actual buildings. Such projects were typologically very different, although they were conceived from single Binishells. This *Binishells spa* promotional catalogue somehow demonstrated that the goodness of this invention lies in its technical aspects, such as the economy achieved from the serial production of concrete shells built from the same formwork, rather than articulated architectural spaces, generated from the intersection of shells of different sizes.

The architectural form which, about fifty years later, characterised Nicolò Bini's residential projects in the United States has been generated from a single, complex,

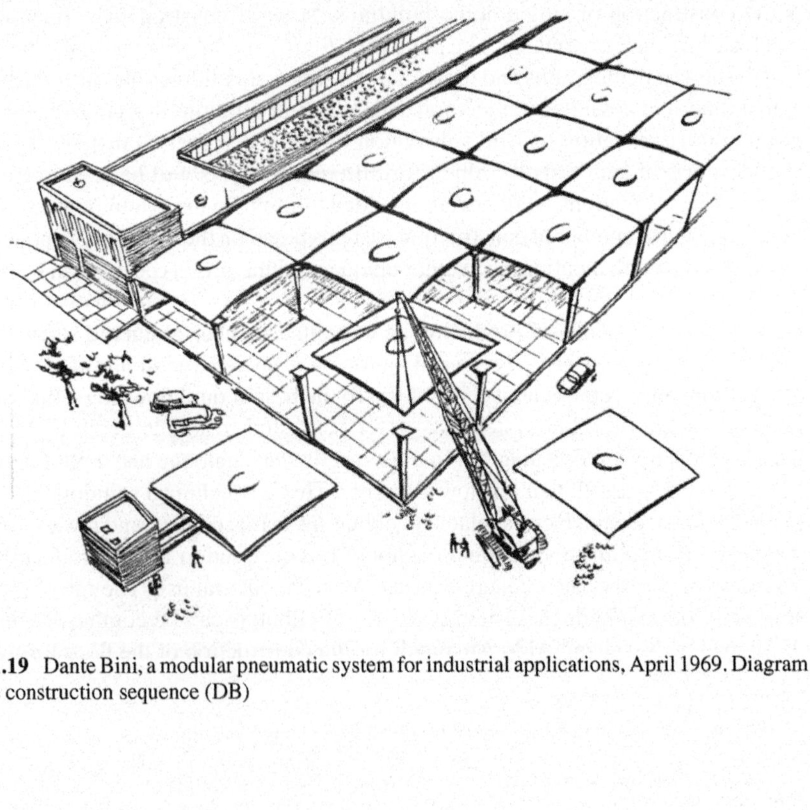

Fig. 1.19 Dante Bini, a modular pneumatic system for industrial applications, April 1969. Diagram of the construction sequence (DB)

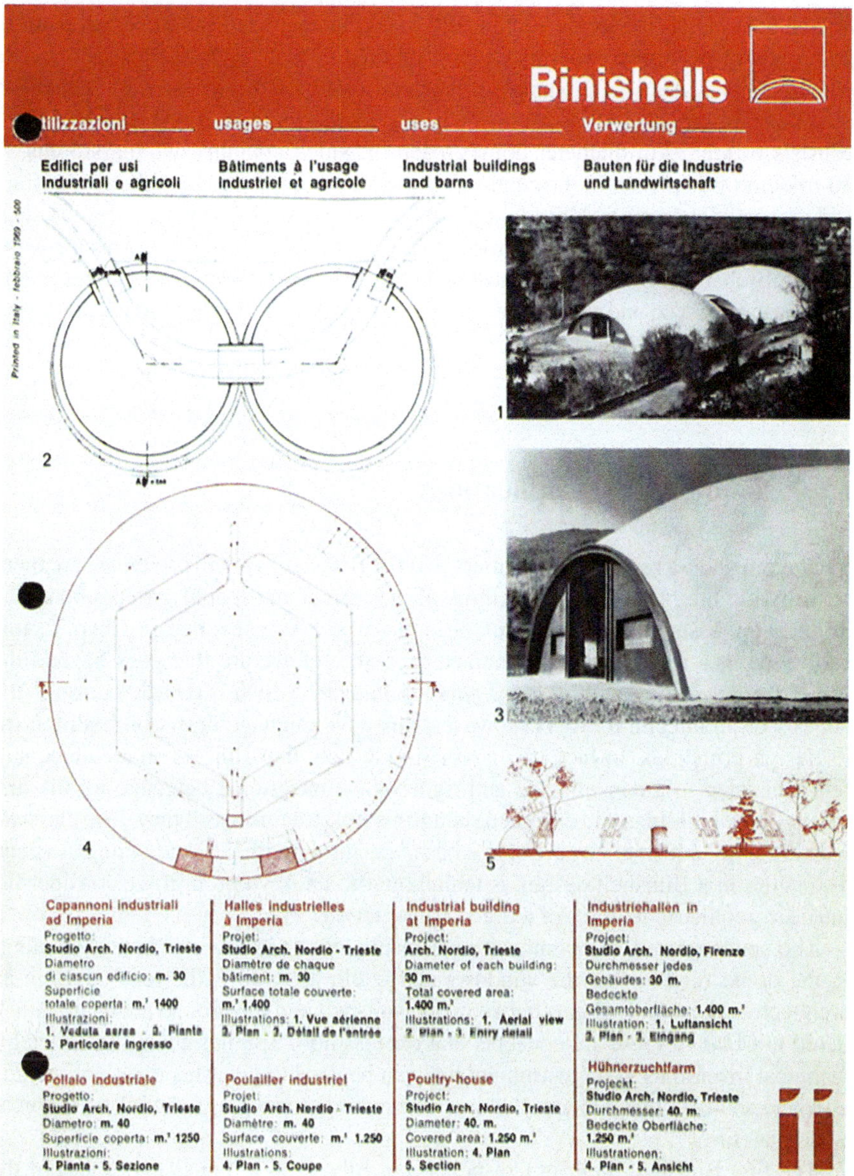

Fig. 1.20 Furio Nordio, industrial Binishell building, Imperia, 1969. Drawings and photographs from a *Binishells spa* catalogue (DB)

ad-hoc pneumatic formwork. Technically speaking, the result is antithetical to those principles of economy and construction rationality that his father had adopted to advertise his systems. An example of such rationality is represented by the design of a school complex by the architect Riccardo Merlo from Bologna. The school consists of ten 20-m diameter domes, each of which contains two classrooms. A 30-m dome was placed in the centre of the school complex to host the gymnasium and management offices. Corridors placed on two separate levels connect the shell units (Fig. 1.21). Two isolated Binishells, 30 and 40 m in diameter, respectively, were duly shaped through large cuts that created glazed façades. The architect Anna d'Alessandris Pazzi and the engineer Angelo Berardi used this system for the design of a motel and a service station (Fig. 1.22). Both projects are analysed later on in Chap. 5.

1.5 Compositional Implications

The layout and compositional frontiers that the Binishell system can create are theoretically infinite; it is sufficient to think of the formal and spatial variety obtainable through the assembly of domes of different sizes. As explained in Chap. 2, this variety defines a language that characterises an architecture that goes beyond the trilithic system advocated by Marc-Antoine Laugier, which is revealed through the genesis of architectural orders: however, this new language draws on tradition for its design principles. In fact, it is interesting to note that Bini, as an architect, still felt connected to a consolidated and rigorous compositional language for his first single-dome buildings and designed such Binishell structures as if they were classical façades divided into three sections: a base, a central body and a crowning element. The latter, in a Binishell design, is tantamount to an oxymoron, if we consider the intrinsic geometric features of a dome construction.

The space-related and compositional results obtained by introducing openings in the dome for housing the windows and oculi, as well as the resolution, in an architectural key, of the joints between the Binishell and the ground are dealt with in detail in Chap. 4: these are concepts that express how Bini has always intended his technical inventions as an instrument that can be used to give shape not only to thin concrete shells to make them effective with regards to structure, but also as a piece of architecture.

The first Binishell, built in Crespellano in July 1964 to house the offices of the Unipack company, is a clear example of this compositional approach. The building includes a 12-m diameter dome and, from a compositional point of view, is based on a clear tripartition: a vertical base, a logical continuation of the ring beam of the foundations, supports the slender structure of the dome. A sequence of identical cuts was made above this base to accommodate prefabricated reinforced concrete frameworks that identify a double row of overlapping isosceles trapezium-shaped openings of two different sizes (Fig. 1.23). The crowing has been designed through the use of a large oculus, which was obtained by trimming the top of the dome.

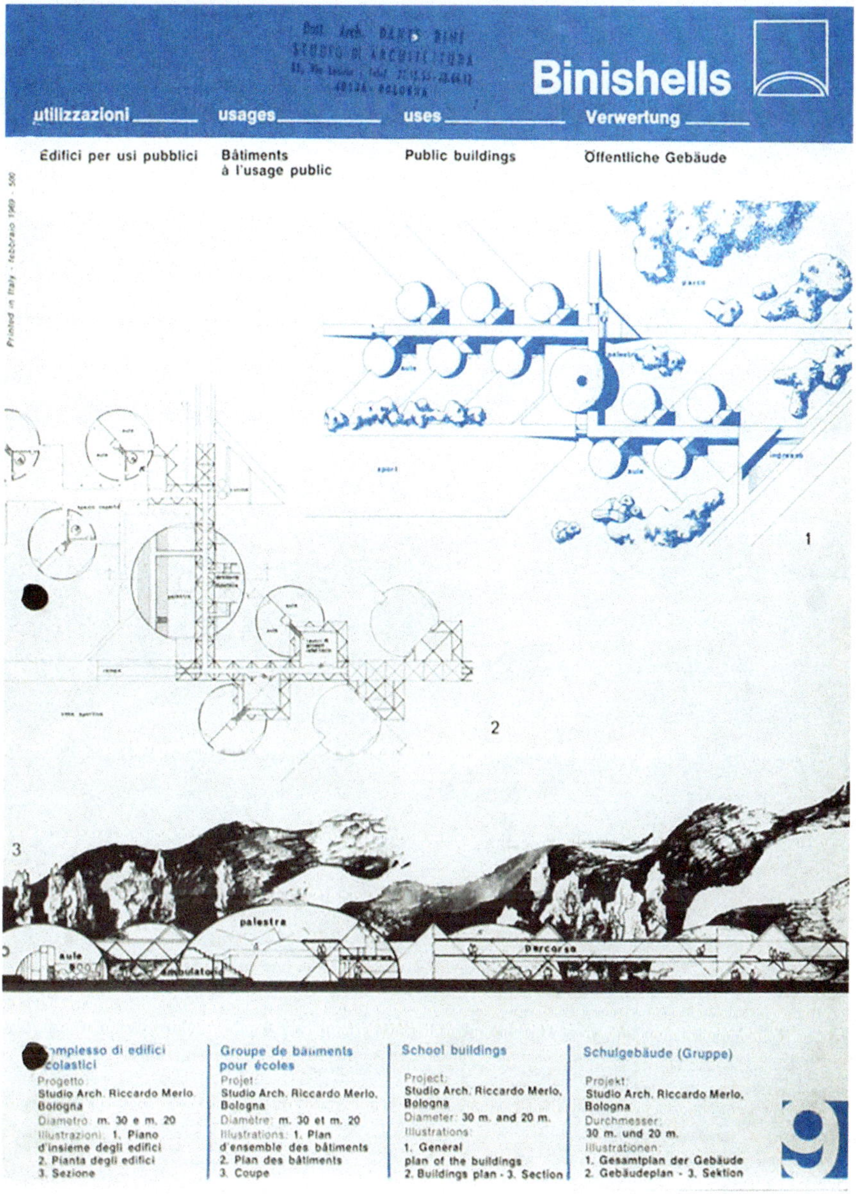

Fig. 1.21 Riccardo Merlo, Binishell school project, February 1969. Plans and section from a *Binishells spa* catalogue (DB)

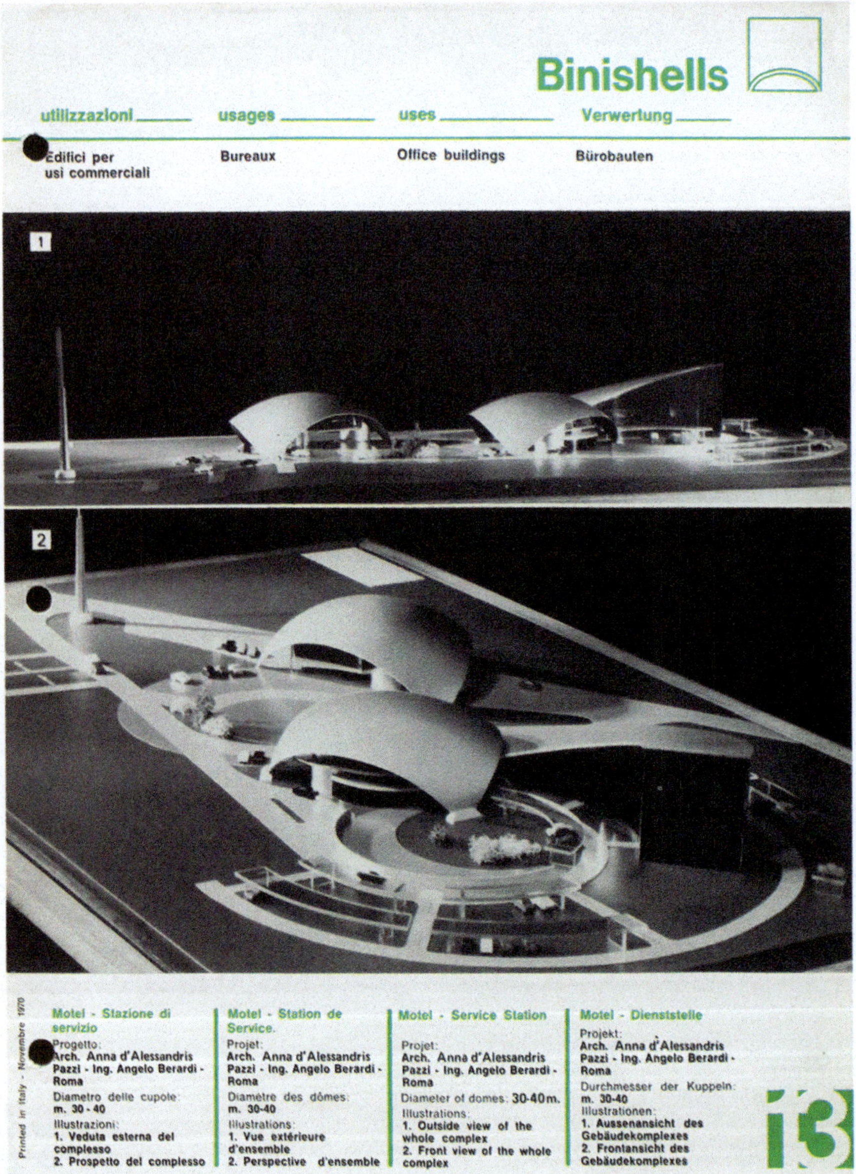

Fig. 1.22 Anna d'Alessandris Pazzi and Angelo Berardi, motel and service station project, November 1970. Photographs of the model from a *Binishells spa* catalogue (DB)

Fig. 1.23 Dante Bini, Crespellano Binishell, Unipack company, July 1964. Detail of the openings under construction (DB)

This attitude towards the design of thin shells was also confirmed in another of Bini's first domes, that is, the 12-m diameter Binishell made in Pegola in 1965. In this case, the base-core system was amplified by using identical, complex and refined vertical prefabricated elements, which were inserted into cuts made in the shell, and to which they were connected by means of arched precast reinforced concrete slabs, assembled after the progressive emptying of the shell sectors between one vertical prefabricated element and the next. At the end of this sophisticated, and certainly costly assembly process, the dome created with the Binishell system was supported by the complex base formed by the assembly of prefabricated elements that formally characterised the building (Fig. 1.24).

This attention to the purely compositional conception of a base can still be seen, for example, in the refined solution of one of the experimental domes built by Bini in his "mushroom field", where the last moulding was used as the framework of the opening to gain access to the building (Fig. 1.25). However, this classical conception of the Binishell would not have guaranteed the economy of construction that was inherent to the invention itself and was progressively abandoned in favour of smooth, polished extrados and intrados shells directly connected to the ground, in continuity with the ring beam of the foundations. Moving beyond a classical compositional approach to dome structures allowed Bini to explore other ways of approaching the architectural expressivity of the Binishell by adapting the geometry of the openings for doors, windows and loggias to the distributive necessities imposed by the building plan or by the surrounding landscape, especially in the case of domestic architecture.

Fig. 1.24 Dante Bini, Pegola Binishell, 1964–1965. Construction sequence (DB)

Fig. 1.25 Dante Bini, an early Binishell prototype at the "mushroom field", San Cesario sul Panaro (DB)

The buildings made using Binishells have only managed to maintain their primordial and distinctive compositional character in a few cases, where the function of the building is not residential and where the need for large windows is evident, as in the case of the thermal swimming pool of the Hotel Ariston Molino Buja in Abano Terme, Italy, built by the Culligan company, where the constructive and compositional solution already experimented in Pegola has been reproposed, although the profile of the dome is an ellipsoid and not a semicircle (Fig. 1.26). A similar compositional principle of a serial repetition of openings of identical geometry, arranged around the perimeter of a dome, was also employed by the architects Michael Godwin and John Faber who, in 1978, used a 36-m diameter and 11-m high Binishell for the construction of the Sports Dome at Malvern Girls' College in Edinburgh: in this case, the designers placed the structure in an artificial lake and created compositional emphasis through the serial repetition of large isosceles trapezoid openings that contain glass windows in direct contact with the water (Fig. 1.27). The base was thus dematerialised and transformed into an abstract element through its identification with the artificial lake: "the final design concept used the Bini structure in architectural rather than purely engineering terms, thus the shell, although entirely raised, was to be cut into sculptural form and placed in a pool, so that all the exterior and interior parts were reflected in the water, and the sunlight reflected over the inside of the dome", Godwin explained (1978: 22).

Godwin and Faber also defined, in their own way, the theme of the base and the relationship of a Binishell with the ground: this is the compositional issue which, together with the geometric arrangement of the openings, identifies the main expressive limits within which designers who use such a construction technology can act to characterise each architecture. This is the case of the Binishell built to house the Impresa Concari offices in Parma, Italy, where the designer chose to raise a concrete shell above a large base, emphasise the presence of the entrance and set the window frame back from the edge of the cupola to emphasise the windows with frameworks (Fig. 1.28). This is similar to a design, which has probably remained unrealised, but which was presented in one of the 1969 *Binishells spa* promotional catalogues, for a sports hall, whose final formal solution would be attained through the intersection of two cupolas. The building was skilfully drawn in section, so as to show a particular relationship with the ground, due to the arched shape of the intrados of the perimeter stands, which created a formal unicum with the extrados of the two shells: their intersection was underlined by an extradosed arch that structurally reinforced the whole system (Fig. 1.29).

It is worth noting how Bini, in his first experimental works, for example in Pegola and Abano Terme, demonstrated a sort of fear when faced with the opportunity of working on a real *façade libre* and experienced the compositional constraint of having to bring everything back to the sequence given by a structural frame that was inserted in a posteriori and artificial way, or to a serial repetition of the various openings: an attitude, as we have seen, that has also been demonstrated by other designers.

The possibility of freely making cuts in the shell without any structural constraints, and of exploiting the homogenous redistribution of compression forces on the rest of the surface, amplifies the concept of *façade libre* beyond the limits imposed by the

Fig. 1.26 Dante Bini and the Culligan construction company, swimming pool Binishell for the Hotel Ariston Molino Buja, Abano Terme, 1968. External view of the prefabricated opening elements (GR)

frame, as demonstrated by the north-east and south-west façades of Le Corbusier's Ville Savoye, where the size of the ribbon windows remains constrained to that of the structural mesh of the entire building. The size and shape of the openings are thus determined by the internal functions, by the layout needs of the building, by the design requirements related to the entry of air and natural light as well as by the architecture/landscape relationship. The building that has probably made Bini famous throughout the world, namely the house designed and built for the Italian film director Michelangelo Antonioni and Monica Vitti in Costa Paradiso di Gallura, Sardinia, conceived between 1969 and 1970 and completed in 1971, is exemplary of this concept. Bini designed a single-dome, 12-m diameter building that houses a single-family dwelling on two levels with double-height spaces. The various rooms overlook the perimeter of the Binishell; in the middle, a patio that opens upwards through an opening in the top of the concrete shell allows air and light to enter the building and contributes to its particular spatial definition, which is characterised by

Fig. 1.27 Michael Godwin and John Faber, Sports Dome Binishell, Malvern Girls' College, Edinburgh, 1978 (MGC)

Fig. 1.28 Impresa Concari building firm, Impresa Concari Binishell offices, Parma, 1969 (DB)

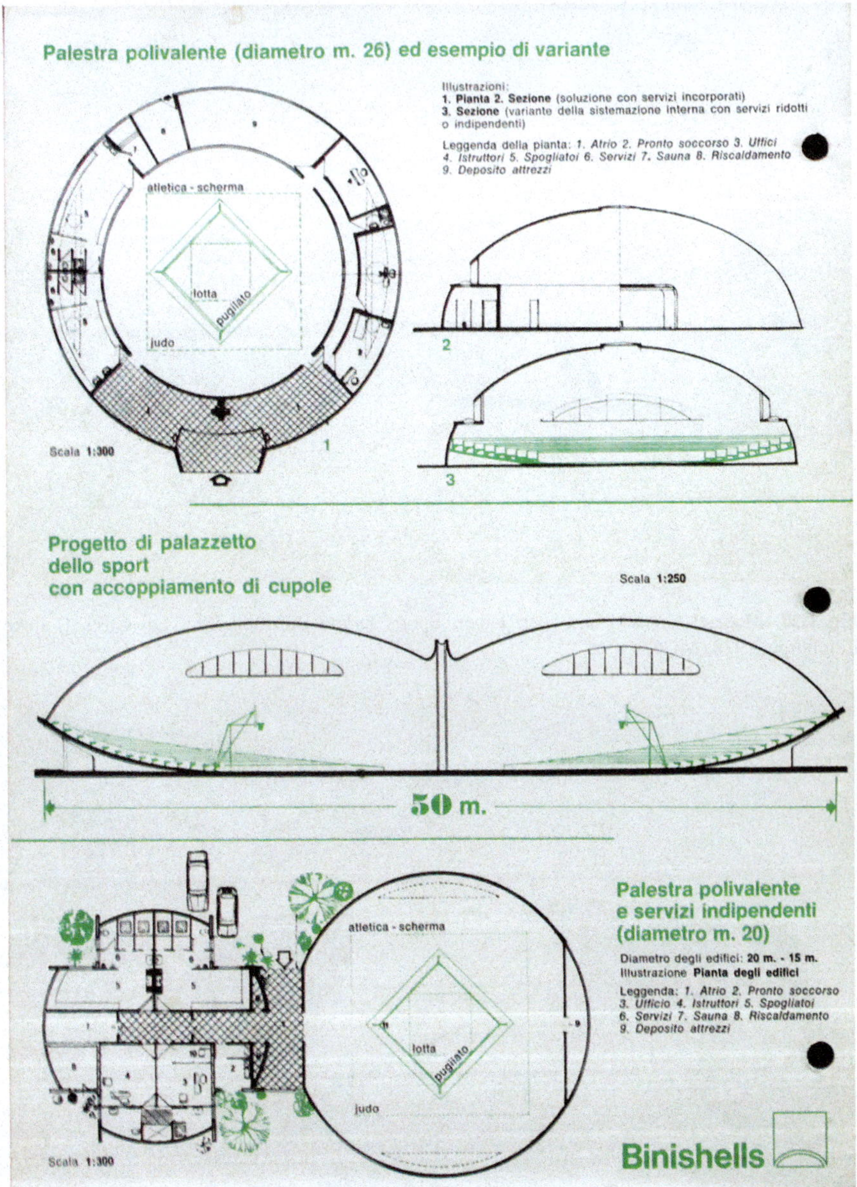

Fig. 1.29 *Binishells spa*, sports hall project, February 1970. Drawings from a *Binishells spa* catalogue (DB)

Fig. 1.30 Dante Bini, Villa Antonioni Binishell, Costa Paradiso di Gallura, 1969–1971. View of the house from above (DL)

irregularly shaped rooms and flights of stairs (Figs. 1.30 and 1.31). All the rooms face onto a central distribution space that can, if necessary, be used as a patio that is opened upwards by means of a large hole that has been made in the concrete shell.

Taking advantage of the morphology of the site, the entrance to the building is via a walkway that leads the visitor to the second floor of the building, where some of the rooms are located. The spatial quality of the interior is attained not so much by the particular shape that has been generated by the shell and the irregular layout of the internal walls, but rather by the system of cuts which opens up the interior space towards the surrounding barren landscape and the sea, so as to give shape to "a shell into which the wind insinuates itself" (Ricci 2016: 138). Bini designed the openings without any preconceived compositional rules, starting only from the requirements derived from the internal layout and the views created from the inside to the outside, and using the most varied shapes and sizes with ease. The free-form shape of the openings contrasts with the regularity of the profile of the dome and thus gives the building that degree of organicity that fits well into the landscape of Gallura.

It is worth noting that the application of the Binishell system in various countries has not affected, from the architectural point of view, the final dome designs in any clearly visible way. It is almost possible to state that the strategies employed to create openings in Binishells and define the overall expressivity of such structures were not influenced to any great extent by the local conditions.

Having definitely abandoned a design approach that was aimed at the tripartition of building or the eurythmy of openings, various designers around the world—and

Fig. 1.31 Dante Bini, Villa Antonioni Binishell, Costa Paradiso di Gallura, 1969–1971. Sea view from the house terrace (AF)

particularly in Australia—have amplified both the concept of *façade libre* and that of *plan libre*, thus expanding the theoretical foundations of architectural design, in the context of buildings conceived from concrete shells.

Dante Bini and many other designers, such as Nordio, Jeleff, Godwin and Faber, have applied Binishells over the years for the construction of circular pavilion buildings, but without ever managing to go beyond the concept of the primitive shelter made up of a single hemispherical shell; in the last decade, Nicolò Bini has succeeded in transforming his father's technical invention concerning the genesis of complex architectural spaces generated by the effective assembly and intersection of sequences of concrete shells made by means of a single pneumatic formwork into a complex geometry and a single continuous concrete casting (Fig. 1.32). Nicolò Bini's main field of experimentation remains domestic architecture: the buildings he has designed and built in the United States show how his design culture is clearly different from that of his father, but from which it is nevertheless unquestionably derived. His spatial approach seems to conceptually follow the layout solution envisaged by Nordio for one of the constructions presented in one of the 1969 *Binishells spa* promotional catalogues, later on discussed in Chap. 5: a large restaurant with a surface area of 880 square metres generated by the aggregation of five concrete shells, each 15 m in diameter.

Dante Bini's invention of the Binishell represents a pragmatic solution that can be used to cover a conceptually simple space, but which is complexified a posteriori, thanks to the cuts made in the shell: this applies to both buildings built with a single dome and to those resulting from the intersection of several domes. Nicolò Bini, on the other hand, has reworked his father's technical invention to provide a constructive

Fig. 1.32 Nicolò Bini, Villa for Robert Downey Jr, Malibu, California, 2012–2017. Model of an early design proposal (NB)

response to a precise idea of space which, from the designer's very first creative impulses, is highly complex and capable of expressing its quality through the spatial sensations derived at a sensory level. It is a compositional and formal evolution that changes the very paradigms of technical invention and the type of client it addresses: as we will see in Chap. 6, Nicolò Bini's masterpiece of domestic architectures is not the result of assembling Binishells of different diameters, but is made from a single concrete casting and required the use of a specifically made pneumatic formwork that was impossible to reuse. With this kind of free-form layout and formal research, there is an explicit renunciation of the standardisation of the construction process, which was one of the main factors that decreed the worldwide success of his father's invention decades ago.

Therefore, Nicolò Bini's formal desire overrides the consolidated constructive rationality preached and implemented by his father Dante, even though his new technique is a revisitation of the original Binishell patent, but which, at the same time, complies with current construction codes. It is a tectonic response to a creative impulse inspired by spatial quality and compositional distinctiveness. Nicolò has questioned the very concept of dome architecture and has developed his designs as sequences of spaces defined by amoeba-shaped shell structures: this is the result of reasoning beyond the constraints of traditional form-found pneumatic domes, which can be serially duplicated, intersected or arranged in a sequence.

According to Nicolò Bini: "this spatial shape is automatically created by the pressure forces within the pneumatic formwork, by the design of the perimeter of the foundation beam and by the material load of the structure. It is no longer invented by the 'whim' of the designer: in fact, it is naturally sized, sculpted and built with absolutely natural proportions by means of an instantaneous self-forming process. In this way, by giving complete freedom only to the forces of pressurised air, three-dimensional and spatial forms are spontaneously created, which are expressive, sinuous, infinite in their variety and extremely efficient in their self-supporting structure; everything can be achieved with a derisory expenditure of energy applied for only a few hours" (2009: 134–136).[16]

[16] Translated from the original Italian text by the authors.

Chapter 2
A New Archetypal Hut in Structural Art

2.1 The Dichotomy Between Dome and Cupola in Architectural Theory

Binishell structures make us reflect on the terminological and semantic dichotomy that is created by the definitions of *dome* and *cupola*, regarding how to interpret the architectural form, whose theoretical meaning was clearly discussed already back in 1960 by Ada Louise Huxtable. She defined the *structural art* (Huxtable 1960: 12 and 16), a concept that was later extended by David P. Billington in 1983 in his book *The Tower and the Bridge. The New Art of Structural Engineering*. "My major objective – wrote Billington – is to define the new art form and to show that, since the late eighteenth century, some engineers have consciously practiced this art, which is parallel to and fully independent of architecture, and that numerous engineering artists have been creating such works in the contemporary world of the late twentieth century" (Billington 1983: 4). The invention of the Binishell and its application to the design of buildings of the most diverse forms, functions and expressions allows us to now develop and update the categorisation introduced by Huxtable and later on by Billington and to link it closely to architecture *tout court*; at the same time, the Binishell offers a response to the widely debated question about what the expressions of construction that can be said to be truly representative of the connection between the architecture and engineering conceived at the turn of the twentieth and twenty-first centuries are. If architecture is interpreted as the art of building, it becomes spontaneous to connect the mannerism of Bini with an idea of structural expression that is not purely the reflection of a technical culture of the architecture project, but is, first and foremost, an act of artistic manifestation.

By offering an interpretation of the terminological dichotomy that exists between *dome* and *cupola* in the thematic context of *structural art*, we are investigating the reasons behind the relationship between form and structure, whose fundamental principles blend with those that govern the statics or structural mechanics of construction.

A. Pugnale and A. Bologna, *Architecture Beyond the Cupola*,
Mathematics and the Built Environment 7,
https://doi.org/10.1007/978-3-031-26735-2_2

When correlating the Binishell with the concept of *structural art*, it is necessary to go beyond the pure meaning derived from static, geometric and structural principles and, therefore, also consider the formal results and the sensory effects derived from the spatial quality generated by the profile—and, sometimes, also from the surface grain—of a given structure. All this should be summed with constructive considerations, particularly with the tectonic concept developed by Gottfried Semper as the art of assembly. Billington himself, when referring to *structural art*, stated that "it is a movement awaiting a vocabulary" (1983: 4). It is also necessary to consider that investigating the construction of architecture projects, if not confined strictly to the discussion of the employed technology and techniques, can become an instrument of pervasive analysis of the built work itself. This aspect is not of secondary importance when the aim of the study is to develop a theoretical and methodological framework for conducting a critical analysis, from a compositional standpoint, of the Binishell architecture.

In such a scenario, it is essential to take into consideration that Bini's invention, to date, is the only strategy conceived to design—or more precisely "find"—a structural form while constructing it at the same time. The Binishell system is in fact able to create a perfect synthesis between two separate challenges and goals in the design and construction of thin concrete shells: the expressivity of structural form through experimentation with form-finding methods and the rationalisation of buildings through the tectonic use of materials and techniques. For this reason, it is worth reflecting in this chapter on the terminological and expressive dichotomy between *dome* and *cupola*, to ensure a more precise use of this lexicon in the critical discussion of Bini's designs.

According to the canonical definition given by Aurelio Muttoni in his seminal volume *The Art of Structures. Introduction to the Functioning of Structures in Architecture*, a *dome* is defined as "a three-dimensional structure with a doubly curved arched form, generated by the rotation of an arch around a vertical axis that passes through the key" (2011: 241); moreover, he specified that the term *dome* should only be used "for roofs with a circular or an elliptical plan" (2011: 95). Bjørn Normann Sandaker, Arne Petter Eggen and Mark Cruvellier proposed a similar explanation in their textbook *The Structural Basis of Architecture*, in which they wrote that "a dome can be thought of as a spatial form that is created by an arch that is spun about a central vertical axis" (2011, p. 350). These two definitions match perfectly with the structural form of Binishells, that is, with their circular or elliptical profiles. However, they exclude other experiments and patents developed by Bini, including the Minishell, which are discussed later on in Chap. 3, from the family of domes. When using the term *shell*, Muttoni referred to "a spatial structure essentially subjected to compression, whose thickness is limited with respect to the other dimensions"; thus "in other words, it is a structure that is similar to a membrane, but under compression" (2011: 105). He then specified that "shells take on the form of a dome when the two main curvatures face downwards. In terms of functioning, this means that two different systems of arch can develop on a surface of this type, each capable of absorbing part of the load" (2011: 106).

Muttoni's reasoning is perfectly aligned with that developed thirty years before by Mario Salvadori in another equally foundational volume and which Muttoni was inspired by, that is, *Why Buildings Stand Up. The Strength of Architecture.* According to Salvadori, "the dome reminds us of a series of identical arches set around a circular base that meet at their top, where they have a common keystone" (1980: 226). In order to explain the structural behaviour of the dome as clearly as possible to his readers, he wrote: "Let us then ignore the minor differences of shape assumed by the dome over its historical development and think of it as a perfect half sphere of a thickness small relative to its span" (1980: 226). Salvadori's definition of dome is narrower than Muttoni's, and it would prevent many Binishells developed with an elliptical profile from being considered a dome, in the canonical sense. In fact, Salvadori's definition would exclude most of them, with the exception of the early prototypes built at the "mushroom field". However, Salvadori's primary concern was not of a terminological nature but was instead related to preparing his personal history of architecture, which he viewed through the evolution of structural systems or configurations, and the works of designers of greater or lesser renown. For this reason, he decided to include the Binishell system in his book.

In *Why Buildings Stand Up*, we find, from the very first pages, that the focus of the narrative is architecture rather than structural design. Salvadori describes architecture as "a young art that had its beginnings only 10,000 years ago when men and women, having discovered agriculture and husbandry, were able to give up roaming the surface of the earth in search of food" (1980: 17). From this premise, he concentrated on the physical, mechanical and structural principles that came to be used, in ensuing centuries, to build shelters (roofs) and crossings (bridges), with a narrative that, although not always chronological, serves as a step-by-step guide to constructive typologies, physical principles, entities such as "loads" and concepts such as "resistance by form". "Form-resistant structures" are thus the subject of one of the most important chapters in the book, and this concept literally fascinated Salvadori until his death. In the chapter dedicated to houses, he revealed his passion for buildings with circular plans topped by domes: from the Sardinian nuraghe, surmounted by corbelled domes, to the stone trulli of Apulia (Todisco et al. 2017), to the wooden roofs that characterise the constructive traditions in the Amazonian jungles of South America, to Eskimo igloos built of blocks of ice and, finally, taking a great leap forward in time, to the prefabricated systems in reinforced concrete used by Nervi in the early 1940s to build his famous hangars for the military. By developing a reasoning based on structural type, Salvadori was able to juxtapose, for instance, the thin shells in reinforced concrete of Félix Candela with the structural systems of Gothic cathedrals, or the physical nature of other well-known domes.

It is not by chance, when describing the spatial effects of such structures as the Pantheon, Hagia Sophia, Santa Maria del Fiore, Saint Peter's, Ahmet's Mosque, Saint Paul's, as well as the Reliant Astrodome, the Louisiana Superdome and the Pontiac Silverdome Stadium, that Salvadori's writing reached the highest level of lyricism: "these structures are the most perfect examples of spatial geometry, whether realised in stone, brick, concrete or steel. No master builder's achievement has attracted humanity as has this most perfect of all shapes", wrote Salvadori, when wondering about the reasons for the great success of this kind of structure throughout the history of architecture. "Is it because of its Platonic purity? Is it because of the separation it erects between the outer, undefined space and its well-defined enclosure? Is it because of the pious or joyous feeling of fraternity it elicits from the participants in its rituals? Or because of the equality created among the throngs? Perhaps the dome is the nearest materialisation of heaven, the only man-made representation of the sky, and this is why a dome seems to protect us like the sky in a clear night, embracing us and our smallness and solitude" (Salvadori 1980: 225). According to Salvadori, "the perfect dome has no scale, no frame of reference", and it is fact this particular feature that generates spatial quality: "when small, it feels limitless, when large, it may embrace us like a room" (1980: 226). Despite these premises, Salvadori, nevertheless, concentrated his explanations on "the dome as a structure", thus confirming the structural and constructive value of the term *dome*, with which *cupola* should never be placed together.

It is important to underline that Salvadori wrote *Why Buildings Stand Up*, which can be considered his personal history of the structural form in architecture, with a narrative that links basic structural forms and principles of resistance by form to a final exegesis on domes, dedicating an entire chapter to form-resistant structures, followed by another two chapters on domes and the case study of Hagia Sophia. After four chapters on loads, materials and historical structures, such as the pyramids, Salvadori's book introduces frame structures—columns and beams. It is interesting to notice that, for Salvadori, a flat slab can easily morph into a form-resistant structure by increasing its strength through simple operations, such as the addition of ribs or a change in shape. When suitably curved, in keeping with one of the three categories defined at the beginning of the nineteenth century by Karl Friedrich Gauss, that is, "domelike", "cylinderlike" and "saddlelike", a flat slab can drastically increase its strength through basic geometric variations. With this "fluid" transition between structural typologies in mind, Salvadori described and interpreted various famous shell structures, such as that of Candela's Cosmic Rays Laboratory in Mexico City. In the book, ample space was given, not necessarily following a chronological narrative, to hyperbolic paraboloid roof structures in reinforced concrete, including those used by Pier Luigi Nervi for the Cathedral of St. Mary in San Francisco and the umbrella-like structural elements in Newark International Airport. There is also a lengthy discussion on masonry vaults and thin shells, which are characteristic of the constructive tradition in Catalonia (used, e.g. by Antoni Gaudí in 1909 for the roof of the school at the Sagrada Familia in Barcelona). These are illustrated according to the procedure patented in the United States in 1885 by Rafael Gustavino.

In the sections about such acclaimed designers as Candela and Nervi, Salvadori, curiously, dedicated a few lines to introducing a method for building shells in reinforced concrete by means of a pneumatic formwork, a technology which, as he explained, was introduced to reduce the cost of the traditional wooden formwork: "Pneumatic forms were first used in the 1940s by Wallace Neff, who sprayed concrete onto them with a spray gun" (Salvadori 1980: 203). This system was later improved by an Italian architect whose fame, in 1980, had not yet reached the same level as the other previously mentioned designers: Dante Bini (Fig. 2.1). Salvadori wrote,

Fig. 2.1 Dante Bini and Mario Salvadori photographed at Villa Bagnaia, Arezzo, in the summer of 1996 (DB)

"Dante Bini sets the reinforcement and pours the concrete onto uninflated plastic balloons, and then lifts them by means of air pressure. The Bini procedure … has met with success over almost the whole world for the erection of round domes of large diameter (up to 300 feet) for schools, gymnasiums, and halls"; but given that "… balloons are naturally efficient when round" he underlined one of the obvious limits by concluding that "these procedures cannot be adapted for application to other thin-shell shapes" (Salvadori 1980: 203).

Salvadori described and considered thin shells, very reductively, as a technical expedient for roofing spaces, and his assessment did not include any specific mention of the aesthetics that could have led certain designers to prefer one solution over another, or of the spatial effects that such structures can generate. This aspect is indicative of how much, when reflecting critically on the evolution of the relationship between construction systems and architectural design in thin shell structures, it is necessary to take into consideration a group of architects and engineers, whose projects of *structural art* have unquestionably written an alternative story of modern architecture that has developed at the same time as the application of Le Corbusier's *Dom-Ino* trilithic skeleton system. This phenomenon is still relatively unexplored, but it is in fact closely related to many contemporary research projects that specifically investigate the relationship between ideation and implementation in architectural and structural design. In this context, the set of projects that authors such as Huxtable and Billington referred to as *structural art* can offer inspiration to achieve an expressive synthesis that blends the principles of construction rationality, automation in construction and form-resistant structural systems. The development of the Binishell patent exemplifies the modern and contemporary design cultures related to *structural art* and stimulates considerations of a compositional and aesthetic nature, on the basis of purely technical factors.

It is therefore no coincidence that Bini himself refers to many of his projects as *cupolas*, resorting to a term derived from the Latin *cupŭla*—which is a diminutive of *cupa*, that is, "barrel", "bowl" or "cup"—immediately bringing us back to an Italian tradition that sees the conception and construction of architecture as a process that blends structural performance with the tectonic use of materials and the achievement of a specific spatial quality of the interiors.

2.2 The Binishell System as a Design Tool

As emerges from the exegesis on the structural form elaborated by Salvadori in *Why Buildings Stand Up*, on the one hand, Bini's result was revolutionary; on the other hand, there is no doubt that it can be considered as the natural consequence of the stimulus provided by the inspiring context he was living in, namely the incredible media success of reinforced concrete shells and form-resistant structures in general (Pugnale and Bologna 2015).

Several designers of the so-called *structural art* were periodically celebrated in the 1960s. For instance, Harvard University conferred the *Charles Elliot Norton Professorship of Poetry* to Félix Candela, Richard Buckminster Fuller and to Pier Luigi Nervi in 1962. They became the poets who represented a new architectural language which was derived either from classical mathematical principles or from basic structural intuitions (Bologna 2013: 31–43). Another example is that of the *Twentieth Century Engineering* exhibition, which was organised by the Museum of Modern Art in New York (MoMA) in 1964, exactly while Bini was erecting his first Binishell dome in Crespellano. The works of many structural engineers were clearly considered at that international event and were exhibited as true pieces of art. Silos, tanks, cooling towers and industrial/service buildings were all characterised by structures of a great iconic impact and were all presented as new monuments of the contemporary age. The most relevant concrete domes presented on that occasion were the Lowry Air Force Base in Denver, Colorado (Fig. 2.2), the Sewage treatment plant in Hibbing, Minnesota (Fig. 2.3) and Nervi's Palazzetto dello Sport in Rome. The curator of the exhibition, Arthur Drexler, considered shells that had been built through inflation as a separate set of experimentations, and two of such projects were included: a 1960 concept by Frei Otto for a complex pneumatic structure in Chicago and the exhibition building designed by the architect Victor A. Lundy for the Atomic Energy Commission (Fig. 2.4). It is important to underline that Drexler did not take into account any examples of reinforced concrete shell-based residential architecture, despite the great success that Wallace Neff's houses had already had at that time: a striking sign of how, in that period, domestic architecture was culturally intended as a design field on its own, still conceptually far from the constructive and formal experimentations developed by engineers.

1964 is also the year the proceedings of the *World Conference on Shell Structures*, which had been held in San Francisco in October 1962, were published. Papers were presented by the most important architects and engineers who engaged in the

Fig. 2.2 Daniel-Mann-Johnson-Mendenhall Associates, Rust Engineering Company and Leo A. Daly, Lowry Air Force Base, Denver, Colorado, 1961 (MoMA)

development and construction of reinforced concrete shell structures, such as Félix Candela, Anton Tedesko and Nicolas Esquillan, as well as those figures who resorted to other materials and typologies, namely Ildefonso Sánchez del Río, Eladio Dieste and Frei Otto.

Salvadori, who attended the conference (*Proceedings* 1964: 688), was completely astonished to discover Bini's invention, which was the perfect synthesis of three innovations in construction technique that had developed independently in previous years: first, the in situ reinforced concrete shell technology; second, the inflatable and pneumatic membrane technology for air structures; and third, the self-shaping steel reinforcing technology (Bini 2014: 37–41).

As pointed out in Chap. 1, the use of pneumatic formworks to build concrete shells stems from Norman W. Mohr's intuition, in 1927, and had already been successfully developed by Wallace Neff during the 1930s and 1940s, as well as by Eliot Noyes for his famous houses built during the 1950s (Pugnale and Bologna 2014).However, the revolutionary concept of assembling steel reinforcements on the ground, and subsequently erecting and bending them to achieve the desired shape, is related to another experimental construction system invented by James H. Marsh,

Fig. 2.3 J. C. Taylor and Anton Tedesko, Sewage treatment plant, Hibbing, Minnesota, 1939 (MoMA)

Fig. 2.4 Victor A. Lundy, Inflatable exhibition building for the Atomic Energy Commission, 1960 (MoMA)

and which was presented at a congress held in San Francisco under the name of the "Lift-Shape" process (Marsh 1964). Marsh underlined that such a method was not "presented as a new concept for the design of shells but rather as a logical construction method that offers comparative speed, safety and economy, with a minimum need for prefabricated components" (1964: 447). The steel reinforcement of Marsh's domes was formed by a grid of bars to which a lightweight mesh was linked. This metallic skeleton was then raised by a crane and temporarily supported in position by wooden poles. Finally, concrete was sprayed to obtain a thin shell, which, in principle, was similar to Nervi's first applications of ferrocement (Gargiani and Bologna 2016: 139–171): the shapes that could be obtained with this technique resulted to be completely free and independent of the construction limits imposed by the need to cast the conglomerate above whatsoever formwork, whether wooden or pneumatic (Figs. 2.5, 2.6, 2.7, 2.8, 2.9 and 2.10).

At that time, Bini was just a clever young architect and builder who foresaw the potential of an architectural system generated through a construction method—which had been the synthesis of his studies on the work of such architects as Adalberto Libera and Félix Candela, as well as of such engineers as Isler, Nervi, Fuller and Otto. Bini stated he had met those figures during his training years, and, on those occasions, he had the possibility of talking about the building systems they had used to construct their most celebrated architectures. Bini has also revealed that "instead of simply copying" such already well-known building techniques, he "tried to find any faults in these construction methods and improve them" with his own ideas: "even though reinforced concrete domes and thin shell structures" amazed him, he "could not accept that the temporary timber or steel formworks used to obtain these

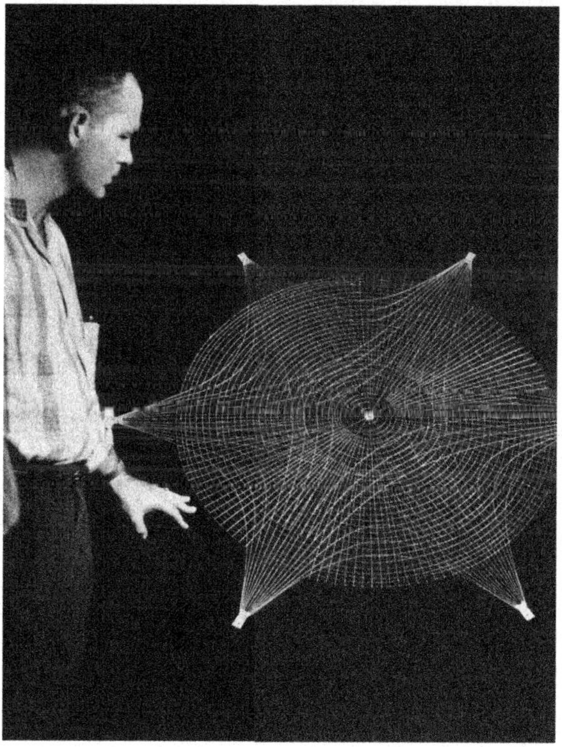

Fig. 2.5 James H. Marsh, "Lift-Shape" process, 1962. Pattern model of a 50-feet span test structure (Reproduced from Marsh 1964: 448)

Fig. 2.6 James H. Marsh, "Lift-Shape" test structure at the Agricultural and Mechanical College of Texas, College Station, Texas, 1962. Lifting and bending the reinforcement on-site (Reproduced from Marsh 1964: 449)

Fig. 2.7 James H. Marsh, "Lift-Shape" test structure at the Agricultural and Mechanical College of Texas, College Station, Texas, 1962. Placement of temporary props after the erection of the steel reinforcement (Reproduced from Marsh 1964: 449)

Fig. 2.8 James H. Marsh, "Lift-Shape" test structure at the Agricultural and Mechanical College of Texas, College Station, Texas, 1962. Concrete coating on the steel reinforcement (Reproduced from Marsh 1964: 451)

sophisticated architectural and artistic engineering expressions, would have cost more than the final structure" (Bini 2014: 26).

As a result, Bini's merit not only consists in having developed a construction technique which, as explained by Salvadori, has achieved success worldwide, but also and mainly, in having generated new architectural formulae, as Bini himself defined them, destined to determine a new formulation of contemporary architecture, even in the residential sector (Bini 2009: 26).

In the previously mentioned 1962 conference, the only presented example of shell-covered residential architecture was that of the Monsanto House of the Future at Disneyland, in California (*Proceedings* 1964: 137–139), which was designed by the architects Marvin Goody and Richard Hamilton and built in 1957 using a fibreglass reinforced plastic sandwich. In this panorama, Bini's invention started animating a completely new research area that did not focus on proposing new housing types based on known form-resistant structural typologies, but rather on the large-scale diffusion of stylistic elements derived from the construction technique he had patented.

The first two domes that Bini erected in Crespellano and Pegola, in Italy, clearly demonstrate how he, as an architect, was aware of the expressive potential of his construction method, already back in the early prototyping stage. Bini's understanding of forces and the structural behaviour of dome constructions translated into a set of design operations that allowed him to transform the enclosed space of a

Fig. 2.9 James H. Marsh, "Lift-Shape" test structure at the Agricultural and Mechanical College of Texas, College Station, Texas, 1962. Concrete coating completion (Reproduced from Marsh 1964: 451)

Fig. 2.10 James H. Marsh, "Lift-Shape" test structure at the Agricultural and Mechanical College of Texas, College Station, Texas, 1962. The finished prototype (Reproduced from Marsh 1964: 449)

newly erected Binishell into a piece of architecture. In most cases, Bini's choices were characterised by a purely decorativist taste that had little to do with the formal engineering research of that period. Bini rapidly developed his stylistic signature through the exploration of possible manners to puncture his shells. He started with cuttings at the base and he then moved on, probably for budget reasons, to other geometrical shapes for openings, windows and intersections between shells. Bini's architecture subconsciously reinterprets the concepts of the *new sensualism* (or *stereo-structural sensualism*), which was theorised by Thomas H. Creighton in 1959 (Creighton 1959a, b). The "sensuous plasticity" of Binishells makes them appear as nostalgic revivals of the many formal experimentations which remained pure utopia, with no practical or commercial outcome.

The house that Bini built in 1969 for film director Michelangelo Antonioni in Costa Paradiso di Gallura, Sardinia, was an ideological synthesis of Neff's 1941 bubble house in Falls Church, Virginia, Friederick J. Kiesler's 1950 sculptural prototype of the Endless House (*The Architectural Forum* 1950) and John M. Johansen's visionary "spray structures" of the mid-50s (*The Architectural Forum* 1959). It was also a pragmatic synthesis that only an architect-builder who was capable of inventing and exporting both architectural stylistic features and effective construction methods throughout the world could realise. It was a business system which, at that time, had already proved to be successful in Italy and Japan and which would then be exported to Australia in the 1970s.

2.3 A Shelter Par Excellence

In architecture, the icon that designers associate with the concept of shelter is generally considered a trilithic structure, mainly thanks to the famous cover page elaborated by Charles-Dominique-Joseph Eisen for the 1755 French edition of *Essai sur l'architecture* by Marc-Antoine Laugier (Fig. 2.11).[1]

During the twentieth century, this typological theme of the shelter inspired various generations of designers to reinterpret and modernise the trilithic structure illustrated by Eisen. In this scenario, the Binishell can surely be considered to embody the concept of shelter par excellence, thanks to the constructive pragmatism of Bini's invention and the architectural expression that he determinately sought in the conception of his cupolas from a compositional standpoint.

In Bini's architecture, the single-dome building, i.e. the Binishell, becomes the main typology that is capable of accommodating the most diverse functions—from offices to schools and various types of housing—thanks to its characteristics of rapidity and economy of construction. However, it is worth noting that the main purpose of a Binishell is to cover a large span with a limited amount of material. This concept prevails over the search for internal spatial quality, which Bini seems to

[1] The English edition, entitled *An essay on architecture*, was published by Osborne and Shipton in London in the same year, with a different cover page designed by Samuel Wale.

Fig. 2.11 Charles-Dominique-Joseph Eisen, drawing for the cover of Marc-Antoine Laugier's *Essai sur l'architecture*, 1755 (Reproduced from Laugier 1755)

deliberately renounce by using inexpensive and conventional partition wall systems or, in the case of wealthier clients, to shape the interiors so as to seek a more explicit formal relationship with the profile of the cupola.

Apart from Bini's designs, which, over the years, have demonstrated the potential of assembling different dome structures for the genesis of projects with complex shapes and a refined internal spatiality, such as the Narrabeen North Public School Binishells, the first experiments that Bini carried out in Crespellano, in Italy, as well as the drawings contained in the patents show how this technical invention was mainly conceived with the pragmatic purpose of covering a space in the fastest and cheapest way possible. The compositional and spatial implications of the invention only seem to have been explored by Bini at a later stage, once the actual feasibility of erecting a structurally sound and economically viable concrete shelter in a very short time had been verified. Unknowingly, Bini thus reinterpreted Laugier's concept of the *primitive hut*, by going beyond the traditional trilithic system, which had been the protagonist, over the centuries, of various types of load-bearing frames, from Vitruvius to Laugier and Semper, up to Auguste Perret and Le Corbusier, with the evolution of the concept of the *abri souverain* in the *Dom-Ino* system. It is from this pragmatism of quickly covering a space with a structure that Bini transformed Laugier's frame, derived from the assembly of tree trunks, into a monolithic reinforced concrete dome.

In 1875, Eugène Emmanuel Viollet-le-Duc exemplified the primordial act of covering a space with the image of a hut made up of a set of branches, arranged around the perimeter of an ideal circumference and connected to each other at one end and then raised to form a pseudo-conical shelter (Fig. 2.12). It was then Bini who solved the same functional problem in a conceptually analogous way, by laying an inflatable formwork above a circular footing ring, on the ground, and then erecting a concrete shell through the simple use of air. The exposed concrete of the internal surface of a newly erected Binishell makes Bini's intention explicit: in several of his projects, the imprint of the steel reinforcement and the geometry of the concave areas between the steel rods are visible, thus allowing us the pneumatic construction technique used as a response to the primordial human need to build a shelter to be appreciated. The photographs of the intradoses of the Crespellano (Fig. 2.13), and the Villa Antonioni Binishells (Fig. 2.14), as well as the shots taken by Max Dupain during the construction of the Ashbury Public School Binishell, in Australia (Fig. 2.15), illustrate this concept and are able to reveal the technological evolution of the used reinforcement system.

The imaginative act that was undertaken by Bini is thus equivalent to that of Viollet-le-Duc, as well as that of Semper or Laugier: they too, before moving to the construction or assembly process, clearly defined a footprint that had to be covered. Bini responded to this challenge by focusing on the design of a construction technique, but the tectonic act that resulted from the application of the Binishell system is only the consequence of an imaginative process, and not vice versa, which, in a certain way, is aligned with the famous statement of Étienne-Louis Boullée: "in order to execute, it is first necessary to conceive" (1976: 83). The pragmatic answer to the question of covering a given surface to generate a primordial shelter capable of giving shape to an architectural space has led, over the millennia, to different typological, structural and formal solutions, depending on the time, place and technical and technological options available to the creator of that space. If a historiographically, well-established vision, defended by rigorous methodologies of academic investigation, has identified the trilithic frame as a structural system capable of giving birth to the Modern Movement and nourishing it, a parallel path of evolution exists in architecture linked to masonry vaulted systems and concrete shell structures. This is an approach to form generation that is not usually considered by architectural historians but rather directly by the architects and engineers who use structural principles to inform the design of their shell and spatial structures. As Salvadori demonstrated in his *Why Buildings Stand Up*, the evolution of form-resistant structural typologies can be considered more as the result of a succession of pragmatic technical experiments linked to masonry and concrete construction than as the evolution of a theoretical framework related to the conception of architectural space (Salvadori 1980: 225–258).

Over the centuries, the literary sources used to identify the theoretical assumptions about the evolution of the Modern Movement have mainly been related to the concept of the structural frame. Even though Laugier devoted a large number of pages to

Fig. 2.12 Eugène Emmanuel Viollet-le-Duc, drawing of a hut with a circular base made up of a set of branches, 1875 (reproduced from Viollet-le-Duc 1875: 6)

vaulting systems in his famous *Essai sur l'architecture*, this treatise is commonly extolled almost exclusively because of a dissertation on the primitive hut, which has fuelled the critical debate on the relationships that exist between structural frame and infilling in contemporary architecture (Oechslin 2002; Meninato 2018; Bologna 2019).

While the structural frame was discussed by Laugier from a theoretical perspective, therefore leaving room for interpretation of the primitive hut archetype, vaulted systems are seen as a practical solution to the construction of solid roof structures. Laugier wrote on this topic that vaults "that have a projection from right to left, require new strength in the walls that bear them" (1755: 147)[2]; "hitherto we could

[2] It is important to note that the author of the 1755 English translation of the 1753 original French text by Laugier mistranslated the original French word "les voûtes" (vaults) as "arches".

Fig. 2.13 Dante Bini, Crespellano Binishell, Unipack company, July 1964. Internal view (DB)

Fig. 2.14 Dante Bini, Villa Antonioni Binishell, Costa Paradiso di Gallura, 1969–1971. Imprint of the steel reinforcement on the internal concrete surface (AF)

Fig. 2.15 Dante Bini, Ashbury Public School Binishell, New South Wales, 1977. Internal view of the concrete structure after the deflation of the pneumatic formwork (MD, DB)

find no better means to support them, than the spurs or buttresses which prevent the walls from giving way. We make use of these in this manner for churches which are properly the only edifices, wherein there are great vaults subject both by their burden and by their height to a great [s]welling. These spurs unhappily necessary, render the outside of our churches very disagreeable" (1755: 147–148). It is clear from this last statement that Laugier was not simply concerned about structural implications, but also about the architectural issues that the use of vaulted systems created: "What I have to observe at present on the subject of great vaults or arches is, that we should endeavour to diminish the weight of them as much as possible. For this purpose two means are advantageous. The first is the exactness of the sweep. The second is the mediocrity of the thickness. The exactness of the sweep contributes infinitely to the solidity of the arches [vaults, AN] and facilitates the support of them. Those that have the knowledge of the sweep of arches [vaults, AN] do wonders with a little expense. Not only is it easy for them to execute these arches [vaults, AN] in such a manner surbased, that they are like a true platform, but they find the secret of sustaining in the air great masses of stone without any appearance of an arch [vault, AN]" (1755: 148–149). To support his reasoning, Laugier mentioned, as an example, the structure of the large and monumental staircase at the Prémontré Abbey: "the boldness of which has something frightful. It is owning entirely the knowledge of the sweep" (1755: 149).

Laugier's considerations, which were based on the observation of masonry vaults, clearly had the aim of understanding the potential of a construction system of becoming even thinner, thus freeing architectural and structural form from mass: "We have of late learnt that they make excellent arches [vaults, AN], that have only one brick in thickness. This old invention in certain countries and new to us shows that it is not necessary that an arch [vault, AN] should be thick to make it solid. Let us take advantage of this discovery, and this will always be to diminish the burden" (1755: 149–150). Laugier thus anticipated the design culture of the twentieth century within which Bini has worked: his intention was to define the most efficient constructive way to cover a large span and generate a shelter through the use of reinforced concrete, capable of definitely freeing the design from the constraints imposed by the older masonry structural systems: "The curved laminar roofing, the shells in reinforced concrete, seemed destined to exalt two of the material's characteristics that tended to free one's imagination: its material inconsistency, due to the extreme thinness, and the possibility to produce totally free forms", Anna Maria Zorgno remarked on this topic (Zorgno 1992: 74).

It is the concept of delineating the footprint that has to be covered by a shelter—defining its boundary through a footing ring beam and placing a pneumatic formwork and steel reinforcement on the ground—that conceptually links Laugier's theories on the mythical origin of architecture—specifically in its shelter form—with Bini's invention. Bini responded to the human need for shelter in a technical way, which is the natural consequence of the time and place in which the Binishell system was developed. Since the architectural implications of using such a system are directly related to the employment of a shell structure and a pneumatic erection technique, it is possible to state that projects generated using the Binishell patent are aligned with the technical advancements desired by Laugier in his treatise, although he could only envisage, at the time of writing, the progressive reduction in mass and thickness of a masonry structure.

Like the frame envisaged by Laugier, the Binishell seems to have been created to respond, among other reasons, to the need for housing. A *Binishells spa* promotional catalogue from May 1967 shows how the firm can offer "a technology that allows the roofing for a small house (150–200 square metres), with absolute solidity and durability, to be achieved in just two days, at a cost achievable by no other construction system".[3] As shown in Chap. 3, both Bini himself and the other architects who have used the Binishell system have, over the years, developed various design proposals for single- and multi-family residences of various sizes.

In the domestic architecture field, it is worth mentioning two Binishell projects, which are antithetical to each other in terms of scope and formal solution: the villa that Bini designed and built for Michelangelo Antonioni in Sardinia, between 1969 and 1971, and the Casa Cupoletta (literally: small cupola house), which was advertised in an Italian women's magazine in the 1970s called *Grazia* (Grace) as an economical response to the desire for an independent house with a garden for the middle class (Olivieri 1975). On one hand, Casa Cupoletta (Fig. 2.16) is a conventional dwelling with fairly standard interiors, and on the other, Villa Antonioni is a holiday house that was destined to become an icon of the architecture of the 1900s (Figs. 1.30, 1.31 and 2.17).

Both projects demonstrate that it is mainly the size of the dome that decrees the boundaries within which the architect can explore the spatial qualities of the interiors.

[3] Translated from the Italian: *Binishells spa* promotional catalogue from May 1967: 13 (DB).

Fig. 2.16 Luigi C. Olivieri, Casa Cupoletta, 1975. External view (reproduced from *Grazia* 1975: 26–27)

They also highlight that defining the number and shape of the openings on a shell surface is also an essential factor in the design of the internal spaces, particularly when the project is site-specific, such as the case of Villa Antonioni.

However, there are a number of conceptual differences between these two projects. Casa Cupoletta was conceived as a sort of modular shelter and was not designed with any specific site in mind. This small dome house was likely to adapt to and develop its own identity in any anonymous suburb into which it was placed. In order to become appealing, and therefore sellable as a kit-home project, Casa Cupoletta was not advertised as a pure dome structure; it included an external wooden pergola and a large chimney, whose presence symbolically indicated the domestic function of the building. In a way, this was a contemporary reinterpretation of a Semperian fireplace around which the family could gather, both inside and outside: in Casa Cupoletta, the fireplace was in fact positioned to enclose the chimney pipes of the fireplace, which was located inside the sitting room, and an external barbecue. In Villa Antonioni, "the pure geometrical shape of the villa is set against the haunting yet impressive backdrop of the cliff facing the sea", observed Lucio Fontana (2014: 150): an idea

Fig. 2.17 Dante Bini, Villa Antonioni Binishell, Costa Paradiso di Gallura, 1969–1971. The house and its surrounding landscape (RC)

of a primitive concrete shelter that completes the barren nature of the site, thanks to its form and tectonic qualities. For example, we understand from Fontana that Bini decided to increase the quantity of local granite in the concrete mix used to build Villa Antonioni. "The resulting building is not one that stands out because of its use of materials. The rough surface takes its colour from the rocks: from dawn until dusk, it blends in with the colours of nature and is simply some 'thing' in the distance. A 'thing' that is truly extraordinary", wrote Fontana in this regard (2014: 151).

2.4 Monolithism as a Point of Strength

Considering Bini's invention as an archetype of primitive dwelling is a reflection that is derived from the observation of the pragmatic and rapid erection process that the Binishell patent offers. As mentioned at the beginning of this chapter, the Binishell is, to date, the only invention that applies a pneumatic form-finding technique at a

1:1 scale, combining the search for structural form with the construction of a reinforced concrete shell in a process that does not take place in separate and consecutive phases. All this is achieved through a limited number of site operations that result in monolithic concrete domes that do not require any particular surface finishes after having been inflated. Therefore, the Binishell is a construction technique that intrinsically generates monolithic concrete structures, which do not include openings: doors, windows and skylights are obtained after the erection process has been completed by removing parts of the concrete shell and creating cuts of the desired shapes and dimensions.

In this framework, it is fair to say that the invention of the Binishell has made Bini the last of the great Italian builders of domes, the direct successor of Nervi, Michelangelo, Filippo Brunelleschi and the unknown builder of the Pantheon in Rome. In fact, the monolithic nature of the Binishell makes it conceptually closer to the Pantheon than to the certainly more iconic domes of the past century, e.g. the cupola of the Palazzetto dello Sport in Rome (Pugnale and Bologna 2017). Indeed, the construction technique developed by Bini is certainly the most suitable for the serial production of domes, even with respect to the better-known system that was patented in 1950 by Nervi (Nervi 1950; Howard 1966: 204–233). After the success of the Palazzetto dello Sport in Rome, Nervi's technique was used to realise roofs all over the world (Gargiani and Bologna 2016: 289–313). In Muttoni's description, the roof of the Palazzetto dello Sport: "is composed of a very slender shell of reinforced concrete. [...] the ribbing of the intrados [...] contribute[s] to the stability of the vault. Moreover, in this case the direction of the ribs does not correspond to that of the compression effectively present in the dome (arch-meridians and ring parallels). At the edge of the central opening we can see the ring that has the job of carrying the horizontal thrust of the interrupted arches. At its perimeter the dome is supported [by] 36 Y-shaped trestles, arranged radially and inclined according to the tangent to the dome. In this way, they directly transmit the thrust of the dome to the foundations" (2011: 97).

The Palazzetto dello Sport is the building that best synthesises the design and construction principles put in place by Nervi during his long career in the construction of thin shells: it is an elegant composition of a number of Y-shaped columns cast in place, supporting a shell roof obtained from the assembly of prefabricated elements. Such elements were serially produced on-site, using a series of jigs, moulds and formworks, and then assembled through connecting ribs that were cast in place. Once assembled, the set of elements prefabricated on-site acted as the formwork that was then used to cast the concrete dome of the sports centre in place. The diamond shape of the prefabricated elements remains visible at the intrados, where it offers the spatial qualities and ornamental features that have made this building so famous throughout the world (Figs. 2.18 and 2.19).

With this on-site prefabrication system that Nervi employed to build the Palazzetto dello Sport, the construction costs are only amortised when the production levels are high; thus, a cost–benefit is achieved by reusing the moulds several times in the case of a large-span roof. Such an economy of scale was also achieved by Nervi for the construction of flat slabs, prepared according to models that were tested for the

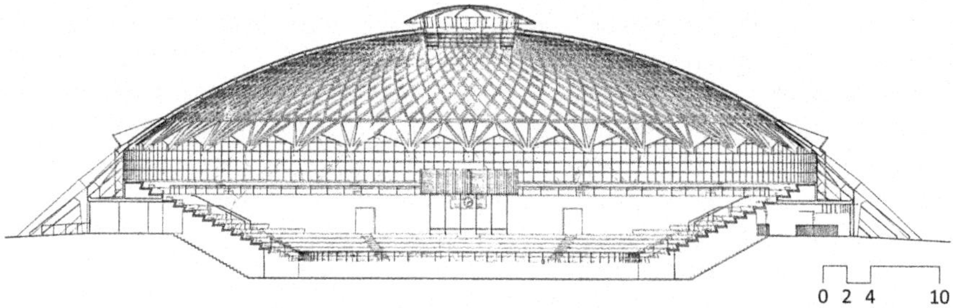

Fig. 2.18 Pier Luigi Nervi and Annibale Vitellozzi, Palazzetto dello Sport, Rome, 1954–1957. Section (AP)

Fig. 2.19 Pier Luigi Nervi and Annibale Vitellozzi, Palazzetto dello Sport, Rome, 1954–1957. View of the ribbed dome intrados (SP)

first time by Nervi's construction company, Nervi & Bartoli, between 1949 and 50, for the building of a warehouse for bales of tobacco at the Tobacco Manufacturing Plant in Bologna. This system was then improved and re-employed for the slabs of the Gatti Wool Factory project in Rome, which also made it famous. The idea is similar and involves the serial reuse of expensive reusable formworks made of thin reinforced concrete elements that are capable of guaranteeing the casting of smooth slab surfaces, without imperfections (Bologna 2015).

The Binishell system originated in Italy in a period in which the media coverage of Nervi's construction systems had reached its peak (Bologna 2013). Although it is natural to consider Bini's invention within this culturally rich period, which led to the development of new and automated construction techniques, there is also a substantial difference between Nervi's and Bini's work. This difference led the two designers/builders to achieve results that, despite starting from similar intentions, can now be considered diametrically opposed, both formally and commercially.

According to Nervi, construction in reinforced concrete should always be regarded as the result of an assembly of parts, and his structures are almost always the offspring of a culture of construction that can be traced back to the tradition of masonry building. Nervi was never able to let go of the concept of a construction that was carried out "by parts", starting from his early systems patented in the 1920s, to those developed for the underground reservoirs and the roofs of airplane hangars of the 1930s, or the project conceived for the E42 arch in Rome, to the patents developed between the end of the 1940s and the beginning of the 1950s for the construction of thin shell structures all over the world up to the 1980s. In fact, there is only a single example in Nervi's work of a shell that was made of concrete cast in place: the dome of the church of San Marcellino in Genoa, which was constructed by his company in 1936 through the use of traditional wooden formworks (Nervi 1965: 102).

Like Nervi, Bini realised early on in his career that reinforced concrete is a liquid material, and true innovation with this construction material could be achieved by improving the design and use of formworks. However, unlike Nervi, Bini was able to detach himself from the masonry-building tradition for the design of his domes: Bini reasoned in terms of monolithicity and not assembly of parts. Furthermore, his method made it possible to almost entirely eliminate the use of temporary equipment on the construction site, with the only significant economic investment dedicated to preparing a reusable inflatable formwork. This meant that labour was reduced to a minimum, as only a restricted number of workers were necessary, workers who were first trained to control the pressure of the framework and then to manoeuvre the mechanical vibrators anchored to the apex of the inflated dome structure. The reduction in the number of workers on-site contributed to the success of Bini's technique and its exportation to other countries, while the strong component of craftsmanship required for the production of the basic prefabricated elements of the domes and vaults designed by Nervi (a process that required highly specialised trained workers) ultimately led to its commercial failure in Italy, and particularly in the United States.

In Bini's work, it is possible to find several cases in which his idea and prototype followed the same constructive pragmatism that led Nervi to design and construct his projects from prefabricated elements. In 1968–1969, in his "mushroom field", Bini prototyped a few concrete experimental structures that included square and triangular prefabricated insulation elements made of a lightweight material (Figs. 2.20, 2.21 and 2.22). These experiments led to the development, later on, of the Binisix (also called Binix) and the Binistar systems (Figs. 2.23 and 2.24). This latter did not even require concrete and was purely made of a structural steel frame. A pneumatic membrane

Fig. 2.20 Dante Bini, experimental Binishell structure at the "mushroom field", San Cesario sul Panaro, 1968–1969. Internal view of the prefabricated insulation elements (GR)

was used to erect this spatial frame and was then kept as the cladding. The Binistar was patented in 1979 in Australia to construct large-span structures—always by means of a pneumatic formwork—made of prefabricated elements of various types and materials, which, in this case, led to the creation of gridshells (Fig. 2.25). Such systems did not have the same success or use as the monolithic Binishell concrete system: as a consequence, Bini did not have the same professional chances to open up to their full expressive and compositional potentiality.

As already mentioned, at the end of the 1960s, the need to amortise the initial cost of the pneumatic formwork led Bini and his company, *Binishells spa*, to design various structures that could be sold using a sort of "kit-home" marketing strategy. Therefore, Bini was attempting a commercial distribution that Nervi could not have done, precisely because of the way his construction systems worked. The various *Binishells spa* promotional catalogues do not simply show Bini's commercial ambitions, and they also illustrate his design research, as derived from an exploration of geometry and space. This allowed other designers to use the Binishell technique and put their own ideas into practice. What followed was a remarkable experiment that was conducted by various architects and engineers, who were capable of designing with the Binishell patent in ways that not only concerned formal expression, but also the programmatic aspects of interior spaces.

Fig. 2.21 Dante Bini, experimental Binishell structure at the "mushroom field", San Cesario sul Panaro, 1968. The triangular insulation elements laid over the steel reinforcement and the pneumatic formwork (DB)

Fig. 2.22 Dante Bini, experimental Binishell structure at the "mushroom field", San Cesario sul Panaro, 1968. Detail of the triangular insulation elements (DB)

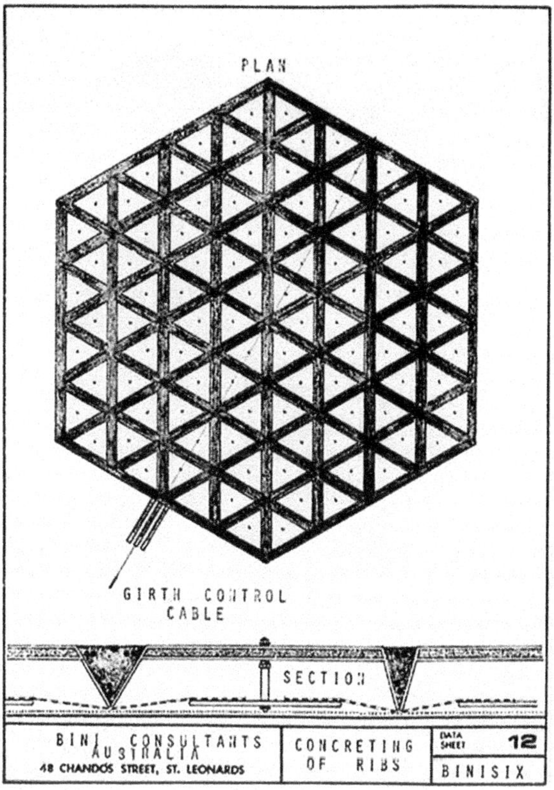

Fig. 2.23 Dante Bini, the Binisix (Binix) system for the construction of lightweight concrete domes, 1976. Plan and section (DB)

In conclusion, the Binishell system represents another form of expression of *structural art* because it perfectly fits the paradigms defined by Billington, who first formulated this concept from the theoretical point of view: "The first fundamental idea of structural art, the discipline of efficiency, is a desire from minimum materials, which results in less weight, less cost, and less visual mass. […] The second fundamental idea of structural art, [is] the discipline of economy, the desire for construction simplicity, ease of maintenance, and a final integrated form. The third fundamental idea of structural art, [is] the search for engineering elegance" (Billington 1983: 266–270).

Fig. 2.24 Dante Bini, the Binisix (Binix) system for the construction of lightweight concrete domes, 1976. Small prototype built in Australia (DB)

Fig. 2.25 Stefano Pietrogrande, a sketch design for a large Binistar sports hall, 1980s (DB)

Chapter 3
Beyond the "Cubic Prison"

3.1 The Binishell is Nonsense

The Binishell system is "nonsense that an engineer's mind could not have conceived". This was Mario Salvadori's way of praising Dante Bini's work in *Why Buildings Stand Up. The Strength of Architecture* (1980: 203). Despite referring to the idea of constructing reinforced-concrete shells simply using air, Salvadori's words could also have applied to the spatial and typological characteristics of most Binishell house designs.

There is nothing strange about a circular building per se, and, on first reading it, this statement could appear paradoxical. The circle has been considered the purest form of art more than once in the architectural theory—from the ancient Greek temples and theatres to the Vitruvian Man, or Andrea Palladio's drawings and Étienne-Louis Boullée's cenotaph, just to mention a few prominent examples. The circle also appears repeatedly in vernacular architecture, and in all those applications that Frei Otto (Otto and Rasch 1995) used to call the "architecture of necessity". There are many reasons why this should not be surprising. For instance, the circular plan maximises the area-to-perimeter ratio of buildings. Moreover, a circular arc provides a decent approximation of a funicular geometry, which is considered structurally ideal because it optimises the use of material. Thus, circles are recurrent forms that can be seen in Igloos, Italian trulli, Indian tepees, Aboriginal shelters, and many other examples of Otto's architecture of necessity.

There is also nothing strange about the idea of designing circular houses, and it is possible to find several examples of dwellings that do not feature a single orthogonal line in their plans. In some instances, a circular house is the result of a purely geometric exercise. In the 1954–56 Mayes House, Don Erickson employed circles of different radii to create the main building volumes, and this choice seems to have had little impact on the final tectonic qualities of the built project (Fig. 3.1). The circle is constantly present in the work of the Australian architect Gregory Burgess, who has

A. Pugnale and A. Bologna, *Architecture Beyond the Cupola*,
Mathematics and the Built Environment 7,
https://doi.org/10.1007/978-3-031-26735-2_3

used this form of geometry to define the proportions of many of his house projects—for instance, in the 1984 Windhover House in Arthurs Seat, Victoria (Fig. 3.2). Burgess has also used the circle for symbolic reasons in various public buildings that he has designed collaboratively with local indigenous communities. The 1990 Brambuk Living Cultural Centre in Halls Gap, Victoria, is a notable example of this kind, which was developed with the participation of the Traditional Owners of Gariwerd.

In other cases, the shape of the plan is the direct consequence of applying a specific construction technique. For example, in Fabrizio Carola's recent masonry domes, the circular or ogival form is an intrinsic feature of the project because it is generated through the use of a giant compass, which is used to place the bricks according to a simple structural design based on geometry (Fig. 3.3) (Alini 2016).In Grilly's experimental house in France, designed in 1959 by Pascal Häusermann, the circular layout represented the logical solution to build a steel structural frame that was later covered by a thin veil of concrete, without using any formworks (Fig. 3.4).

Anna Maria Zorgno wrote that very few designers "managed to confer the dignity of a product of architecture to shell roofing", thus guaranteeing "the confluence of the architectural and volumetric matrix with the constructive and structural matrix"

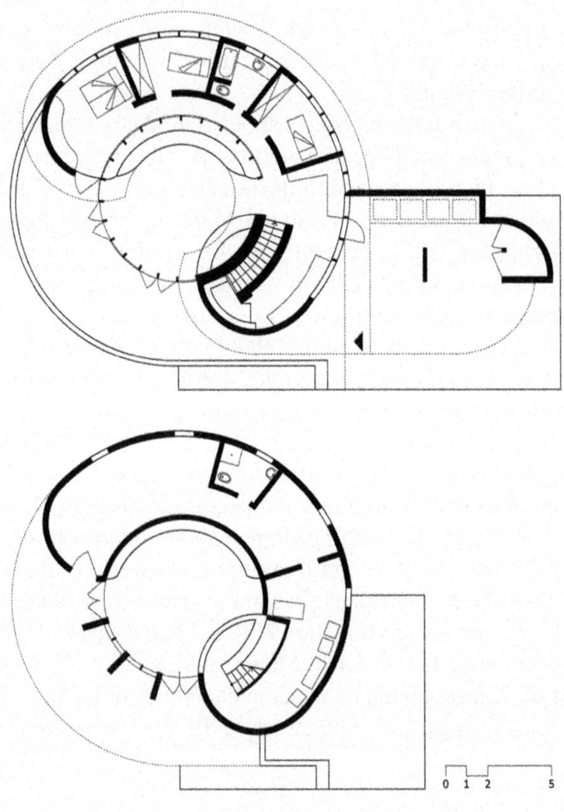

Fig. 3.1 Don Erickson, Mayes House, Glen Ellyn, Illinois, 1954–56. Ground-floor and first-floor plans (AP)

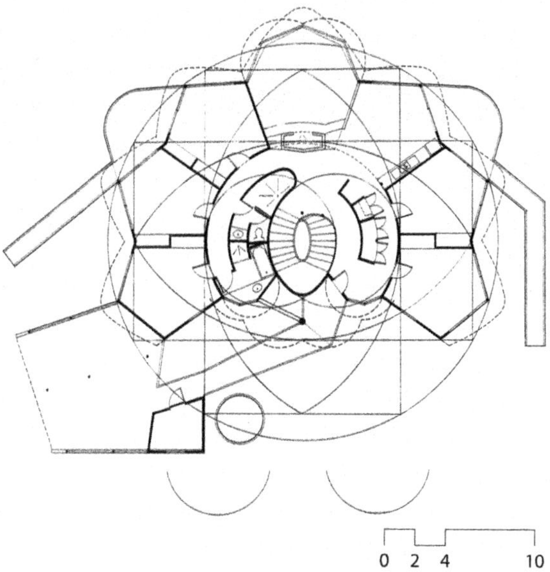

Fig. 3.2 Gregory Burgess, Windhover House, Arthurs Seat, Victoria, 1984. First-floor plan (GBA)

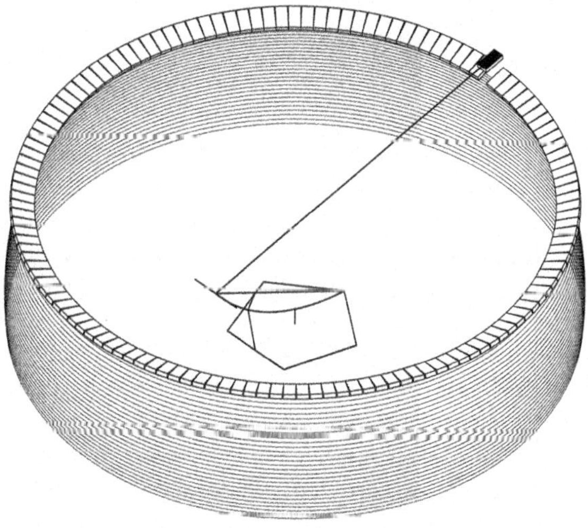

Fig. 3.3 Fabrizio Carola, compass technique. Scheme (AP)

of the building (1992: 75).[1] It is not so relevant here to establish how close or far from reality this statement is. However, it can certainly be said that not all design briefs lend themselves to being developed using shell structures. It is also true that some building typologies provide more opportunities for implementing shell structures

[1] The title of this chapter is a tribute to Zorgno and her scholarly work. See her 1992 journal article entitled: "Beyond the Cubic Prison".

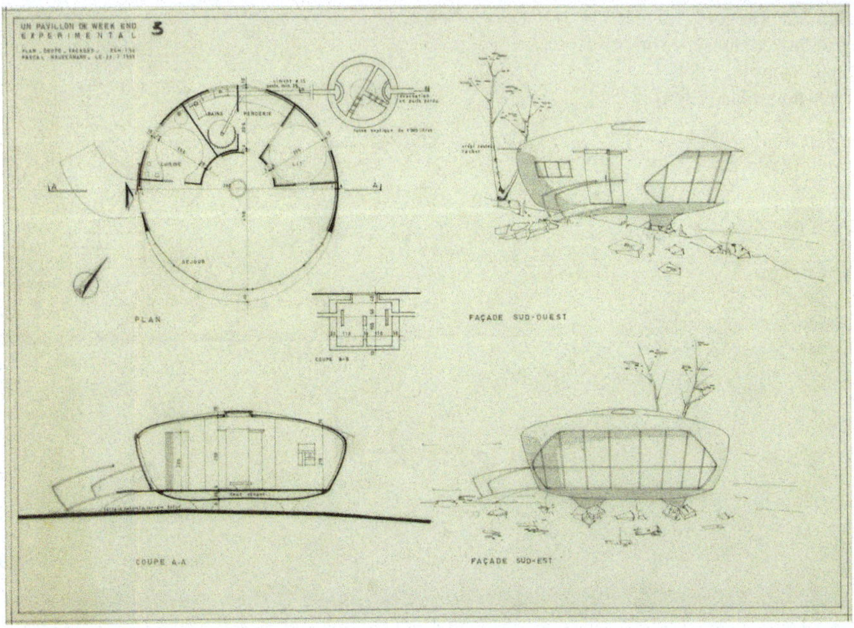

Fig. 3.4 Pascal Häusermann, experimental pavilion, Grilly, 1959. Plan, section and elevation (CFCVL)

than others: it is more common to see large spans in markets, museums and sports centres than in single-family houses and small offices.

Having said this, it is still possible to find a few residential projects that have implemented shell structures. Lloyd Turner's house in Boulder Creek, California, is a curious example that was built in 1982 using pneumatic membranes (Turner 1986). Turner used a soap-bubble cluster as the conceptual construction model, where the whole cluster represented the house, and each bubble identified the individual rooms and functions (Figs. 3.5 and 3.6). This strategy allowed Turner to shape architectural spaces using air from the inside and looking outside, with the overall form and size of the house resulting from and expressing the internal arrangement of the rooms. Turner's approach worked well for a one-off application, but would have certainly shown its limits if standardisation, repetition and low budget had been among the project requirements. The cost of using ad-hoc made pneumatic formworks and concrete reinforcement is not a negligible factor in this type of structure.

The Binishell system guarantees cost savings as a result of the application of standardisation and construction automation principles, but the architect is forced to design the house from the outside to the inside. The building footprint and overall volume are always given by a Binishell—the construction technique does not simply inform; it defines the house form. Architectural spaces can only be conceived within the constraints dictated by the shape and size of a standard pneumatic membrane.

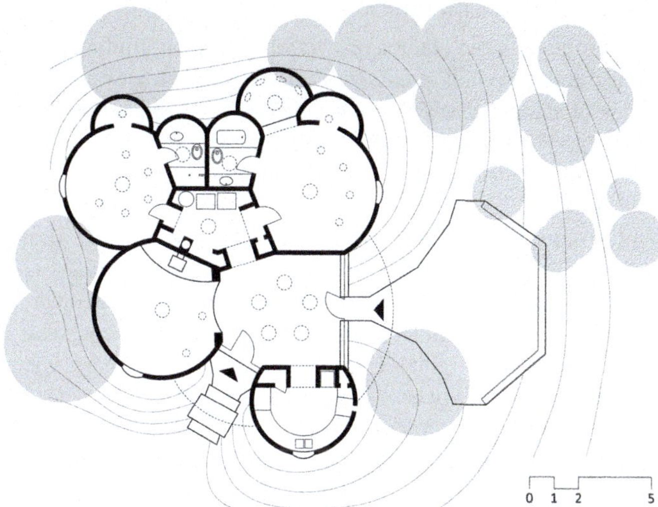

Fig. 3.5 Lloyd Turner, air-formed concrete shell house, Boulder Creek, California, 1982. Plan (AP)

Fig. 3.6 Lloyd Turner, air-formed concrete shell house, Boulder Creek, California, 1982. Construction sequence (LT)

This explains why the early Binishell system can be considered "architectural nonsense". It is in itself quite challenging to match the programmatic requirements of a small house to the spatial qualities of curved shell structures. Imagine having to add the rules of standardisation and modularity to the equation. For this reason, Bini tried to readapt his first Binishell system more than once throughout his career to make it more suitable for small- and medium-scale applications, and particularly for residential projects.

This chapter aims to reconstruct the path that gradually led Bini—primarily through the development of his Minishell and Pack-Home systems—to move away from the circular plan and embrace standardisation, not only in construction but in all the aspects of a house design. However, this narrative only represents one way of tracing the evolution of the original Binishell system over the years. Focusing on the plan and the intrinsic circularity of the first Binishell patent allows us to investigate to what extent construction informed design in Bini's work. At the same time, it also highlights when and how specific design ideas or requirements impacted the development of new construction techniques based on the use of air. This story does not cover several aspects that have also contributed to modifying the original Binishell concept and the definition of new systems. For example, cutting the shell or intersecting domes are operations that deserve separate discussions and are the focus of Chaps. 4 and 5, respectively.

3.2 To Circle, or Not to Circle

Successful applications of concrete shells are easier to find in large-scale projects than in small schools or houses because the "confluence of the architectural and volumetric matrix with the constructive and structural matrix" (Zorgno 1992: 75) of a building is difficult to achieve when an open plan is not an option. The presence of walls, suspended ceilings or any other devices used to subdivide the internal spaces and conceal the services would inevitably interrupt the continuity of a shell roof and its legibility as a structure. Although the "cubic prison" is a practical packaging solution for small cells, shell structures do not like to be confined in bars—they want to enclose large and empty spaces. But this represents only half of the problem and is primarily visible in section or perspective. In the plan, the clash is between the geometry of the partition walls, which usually subdivide the internal spaces orthogonally (Steadman 2006), and the shape of the shell footprint, which is rarely orthogonal.

Images of the first Binishell in Crespellano—erected in 1964 as the Unipack headquarters and demolished in the 1980s—suggest that Bini had been well aware of this design issue right from the beginning of his journey as a designer of form-resistant structures (Figs. 3.7 and 3.8).

There are at least three reasons to believe this. First, the photos clearly show that the internal partitions did not touch the shell. They appear to have been about two metres in height, which was sufficient to define the internal circulation and provide privacy

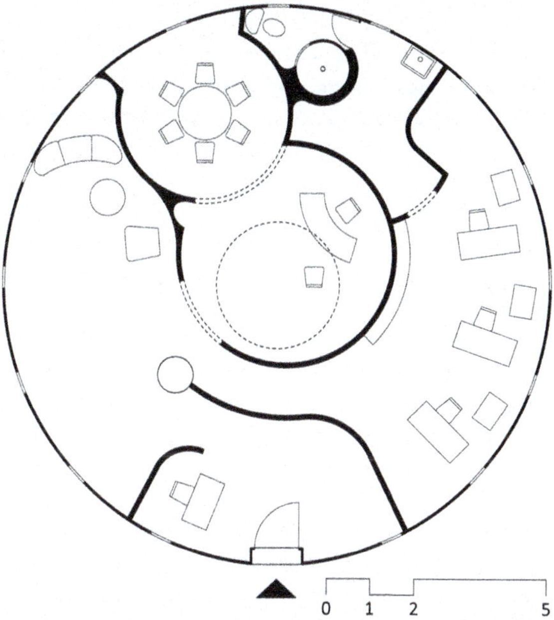

Fig. 3.7 Dante Bini, Crespellano Binishell, Unipack company, July 1964. Plan (AP)

Fig. 3.8 Dante Bini, Crespellano Binishell, Unipack company, July 1964. Internal view (DB)

to the inhabitants of the individual office spaces but not too high to spoil the legibility and experience of the space underneath the Binishell as a single volume. Second, the plan illustrates that the functions were all arranged around a central circular area, which was, for geometric reasons, the largest void inside the building. The importance of this space was further enhanced by the presence of a circular skylight at the apex of the Binishell, which immediately evoked the oculus of the Pantheon in Rome. Third, the shell inner surface was intentionally not rendered or finished in any way. The exposed concrete allowed the imprint of the steel reinforcement and the geometry of the concave areas between the steel rods to emerge, thus providing visual clues about the technique used to build the structure. The uncommonly wrinkled texture of the concrete surface was further proof that the Binishell was shaped using air pressure.

This first example already highlights the three key characteristics that contribute to reinforcing the connection between architecture, structural form and construction technique in a Binishell project. First, it is essential to ensure that the internal space is perceived as a single volume, uninterrupted by visual obstructions. Second, it is vital to consider the role played by the geometry of the cupola while arranging the internal programmes, as the central space of a Binishell is experienced as a much larger void than the areas along its perimeter. Third, the use of a pneumatic formwork should be somehow revealed or expressed by the tectonic qualities of the final building.

Binishells. A new technique for new forms is the title of a *Binishells spa* promotional catalogue printed in May 1967.[2] This catalogue was used to promote the Binishell technology and included a comprehensive description of this patented construction system, images of tests performed on prototypes and a short essay by Salvadori on the technical issues of designing and constructing thin shells.[3] The catalogue also contained many examples of conceptual and built applications of the Binishell technology, sorted according to the size of the dome, and ranging from 12 to 80 m in diameter.[4]

The smallest among these applications is a house project developed in 1966 for the Aluminium Company of America (Alcoa); the prototype was built in Bini's "mushroom field" (Fig. 3.9). At the time, the Executive Director of Alcoa—Anthony J. Faranda—was interested in using the Binishell patent to build houses for those workers they had employed in tropical areas, such as in Liberia where they had bauxite caves.[5] Protecting the residents from the sun and the heat was an essential requirement in such a design, which explains why Bini decided to trim the surface of the shell to create three large terraces sheltered by the concrete structure. Bini used three vertical planes for the cuts and defined a new footprint for the Binishell in

[2] The Italian title of the *Binishells spa* promotional catalogue is: "Binishells. Nuova tecnica per nuove forme". This document is available online at: www.binisystems.com.

[3] The title of Salvadori's article for the Italian *Binishells spa* promotional catalogue is: "Problemi tecnici della progettazione e della costruzione di volte sottili" (Technical issues in the design and construction of thin shells).

[4] The *Binishells spa* promotional catalogue erroneously describes the first small house as a 10-m diameter dome. Bini has confirmed he has never constructed a Binishell that was smaller than 12 m in diameter or 4 m in height.

[5] This story was told by Bini during an exchange of emails with the book authors.

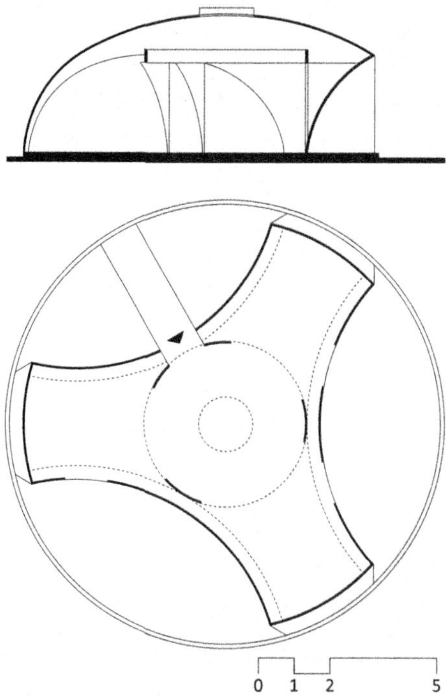

Fig. 3.9 Dante Bini, house project for the Aluminium Company of America, 1966. Plan and section (AP)

the shape of an equilateral triangle inscribed in the original circular perimeter. The internal area of the house was therefore reduced to 55 m^2; enough for a living room, a kitchen, two bedrooms and a tiny bathroom. The kitchen, bedrooms and bathroom were placed in the peripheral areas, close to the shell support points, while the living room was located in the central space, thus connecting all the other functions.

Bini verbally confirmed that the project had not been carried out according to his design, and it was instead built as a pure shelter, without rooms or internal partitions. Despite this, the "Liberian house" remains a good case study because it demonstrates that applying the three principles identified above can strengthen the relationship between architectural form, structural typology and construction technique, even in a simple and small Binishell design. By focusing on the house project as drawn, it can be seen that the living room was conceived to be experienced as an open space, defined by circular partition walls which did not touch the shell. This choice follows the first principle above because the internal space was designed to be perceived as a single uninterrupted volume. In the drawings, the centrality of the living room is emphasised by the presence of a skylight, which confirms Bini also implemented the second principle to merge architectural and structural form. However, the most remarkable characteristic of this small house lies in how Bini managed to express the magic of his pneumatic construction technique through the design of the openings—the threshold between indoors and outdoors. Instead of installing standard window components, he created three rendered concave surfaces as infills. The form of these

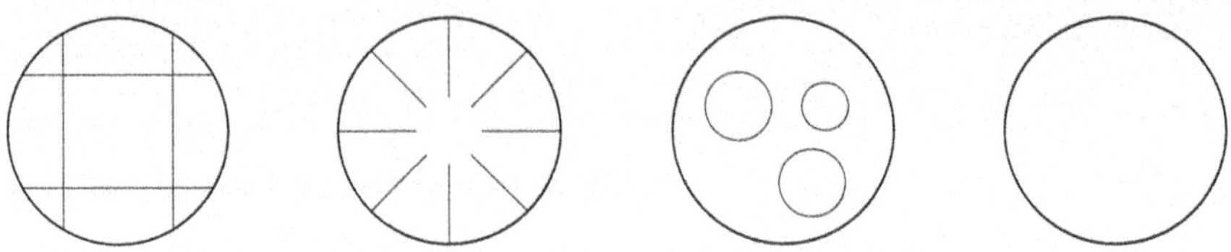

Fig. 3.10 Four ways to organise the interiors in circular buildings, readapted from Anna Szczegielniak (2019). Scheme (AP)

double-curved solid walls was obtained by rotating the arch of the opening about the line of its footprint on the ground plane. The final shape somewhat reproduced the tectonic qualities of an air-formed concrete structure, thereby transforming a non-structural element into an architectural device that conveys an otherwise invisible message.

This approach has at least one famous precedent in Ludwig Mies van der Rohe's Lake Shore Drive Apartments, where Mies could not show the structural frame directly on the skyscraper façade because the fire code required that the steel members should be encased in concrete. To ensure the building skeleton was legible from the outside, Mies decided to cover the concrete-encased columns with steel plates and weld additional I-beams onto those plates to reproduce the original shape of the actual structural elements (Carter 1974).

The "Liberian house" prototype is one of the few examples of residential Binishells that had already been built by the time the 1967 *Binishells spa* promotional catalogue came out. However, the catalogue also featured several unbuilt projects worth considering for the purpose of this chapter. For instance, it included a 20-m diameter experimental house, developed on two storeys and designed by Bini's friend, the architect Furio Nordio.[6]

In *Circular Plans in Contemporary Housing Architecture*, Anna Szczegielniak (2019) identified four different ways of organising the internal space of a circular building: first, with partition walls that are designed on an orthogonal grid, and where only the external walls are curved; second, with partitions that are arranged radially, and which are therefore perpendicular to the curved boundary of the building; third, with the so-called circle-in-circle approach, in which smaller circular spaces are defined within the larger circular footprint of the building; fourth, as an open space, with no internal partition walls (Fig. 3.10).

In his experimental two-storey house, Nordio designed the interiors by mixing the radial arrangement of the rooms with the circle-in-circle approach (Figs. 3.11, 3.12 and 3.13). Even though curved partition walls are present more or less everywhere in the plan, two large circles are predominant. The first circle is located in the centre of the Binishell, thereby identifying the living room on the ground floor and the

[6] The promotional catalogue did not acknowledge the architect's name. Email exchanges with Bini confirmed this experimental house was designed by Nordio, who also worked on other Binishell projects.

children's playroom on the first floor. This circle also defines the most horizontal and vertical circulation spaces inside the house. The second circle encloses the garden area. It is comparable in size with the former and is tangent to the base circumference of the shell roof. Although the outdoor garden is not technically part of the interiors, it visually expands the size of the main functions of the house, by moving the barycentre of the composition from the geometric centre of the dome. The living room, library and bedrooms are placed around the centre, as are the kitchen and the bathrooms. The dining room is the fulcrum of the house at the ground level. Had it not been for the children's playroom, which is also the central space on the first floor—this area could have featured a double height thanks to the form of the shell.

In this project, the possibility of creating a large and open central space is somewhat compromised by the very concept of a two-storey house. The general openness of the interiors also suffers from the spatial arrangement of a dwelling on two levels. The Binishell form is quite legible while moving from one space to the other, but, when inside a room, partition walls prevent it from being experienced as a whole.

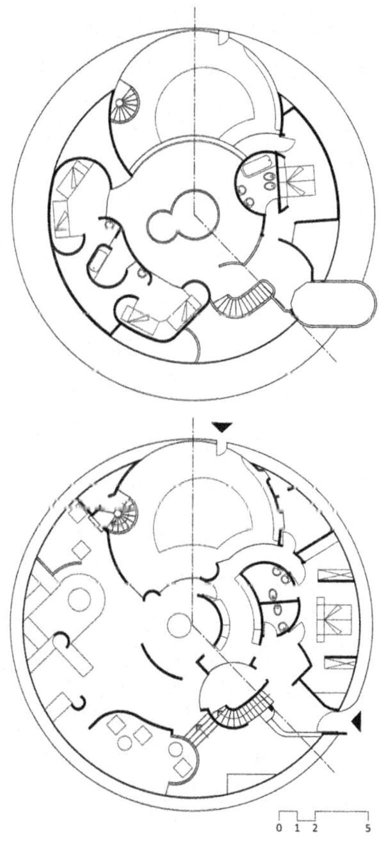

Fig. 3.11 Furio Nordio, experimental Binishell house project, 1967. Ground-floor and first-floor plans (AP)

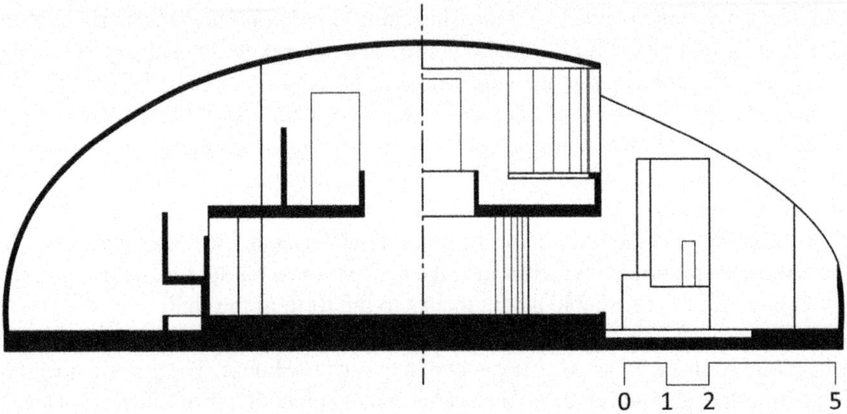

Fig. 3.12 Furio Nordio, experimental Binishell house project, 1967. Section (AP)

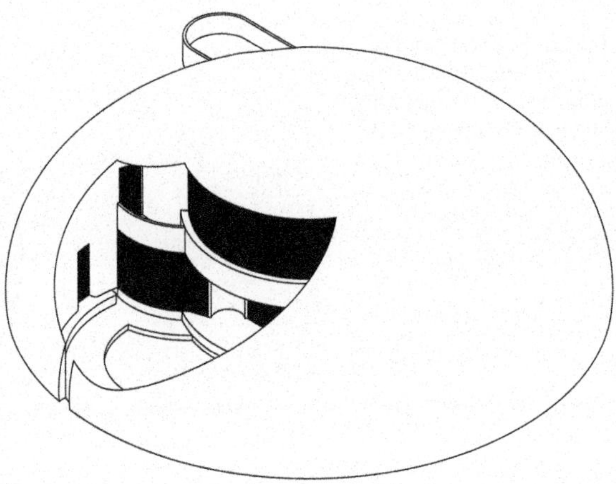

Fig. 3.13 Furio Nordio, experimental Binishell house project, 1967. Axonometric view (AP)

Since this experimental house is an unbuilt project, it would not make much sense to discuss its tectonic qualities or to what extent the final design can reveal details of the construction technique. Nevertheless, it is at least possible to notice how the large opening in the shell appears to result from a simple geometric exercise, i.e. the solid intersection between the subspherical dome and the cylinder generated by the vertical extrusion of the garden surface. The main ambition of such an exercise has probably nothing to do with tectonics; it seems more of a symbolic gesture related to the redundant reproduction of the circular shape in every part of the house.

Analysing the symbolic meaning of this large cut in the Binishell—which could look like the divine eye to some, and a UFO landed on earth to others—would not lead the discussion anywhere. It is far more important to notice how this cut in the structure somehow manages to undermine the very concept of the Binishell system, which intrinsically sees architectural design as subordinate to the construction technique.

In this experimental house, it is as if Nordio started designing from a dome already built on the site. It is as if the Binishell were part of the landscape when he began sketching. Therefore, it appears he decided to shape the house through a process of volume subtraction to demonstrate that removing parts of the structure could generate an architecture that respects the geometry of the Binishell, without becoming a device to enhance its monumentality. Nordio was not afraid to compromise the integrity of the structural form to gain in other aspects, such as the openness of the interiors towards the outside or the liveability of the house.

Architectural design is, after all, an art that is based on compromise, and it is normal to expect that different projects can express the relationship between form, structure and construction technique in contrasting ways. For example, in the Crespellano Binishell, each aspect of the building served to celebrate the cupola. On the contrary, in this experimental house by Nordio, the geometry of the dome was only a starting point, not a goal. It was an impulse that stimulated the architect to break free from Zorgno's "cubic prison" and the standard concepts and forms of the modern dwelling.

Villa Antonioni, designed by Bini himself, is a unique project that privileged integration into the context above everything else (Figs. 1.30 and 2.17). This Binishell sits on a sloping site and acts as a platform to contemplate the surrounding landscape and the sea. Bini met Michelangelo Antonioni and Monica Vitti in 1968 and—for the first time—was commissioned by a private client to design a cupola house. Antonioni had very clear ideas: he knew what he wanted and planned to use this Binishell in Sardinia as a holiday house and filming location. It was not one of those situations where the Binishell company simply licenced out their patent to build dozens of identical structures, all based on the same design. On the contrary, it was a rare occasion to design a Binishell house for a particular site—an untouched natural landscape in Sardinia (Figs. 3.14, 3.15, 3.16 and 3.17).

Here, the combination of an unconventional client with such a unique location led Bini to focus more than he usually did on the house surroundings and external views, somehow reducing the role of the Binishell structure to that of a visual and physical shelter. It is also worth adding that—technically speaking—there was no advantage in using a pneumatic formwork for a one-off application, as construction costs only become competitive when the same design is repeated a number of times.

The central area of the house is an open-air garden and a circulation space; the kitchen, living room, bedrooms and bathrooms are all placed along the perimeter. In terms of pure spatial logic, Villa Antonioni does not differ so much from Frank Lloyd Wright's Glenn McCord House, or even from the ancient Oval House in Crete, in which circulation spaces have also been relegated to the core of the building. The journey through the house begins by crossing an external footbridge that leads to the entrance, which is located on the first floor. The living area can be reached by going down a staircase, while enjoying a framed view of the sea.

A simple way of evaluating the relationship between form, structure and construction technique in Villa Antonioni is to follow the approach used above for the Crespellano Binishell, thus focusing on three main design features: first, the subdivision of the internal space to check whether the cupola can be perceived as a single element

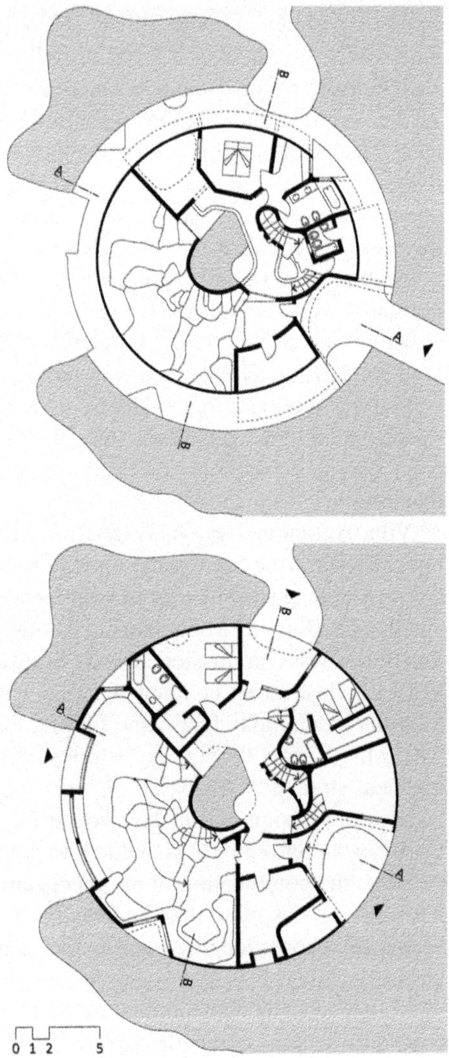

Fig. 3.14 Dante Bini, Villa Antonioni Binishell, Costa Paradiso di Gallura, 1969–1971. Ground-floor and first-floor plans (AP)

or, conversely, was fragmented into smaller parts; second, the role of the space underneath the apex of the dome to determine whether it is central or hierarchically subordinate to the overall composition of the house; and third, the tectonic qualities of the Binishell to verify whether the use of a pneumatic construction technique was expressed in the final building or used to inform its materiality in any way.

Such an approach might sound illogical, since the two projects have very little in common. However, the few aspects they share are fundamental to understand how much Bini weighted architecture against construction in his designs. After all, Bini aimed to build a modern version of the Pantheon with each of his projects, despite

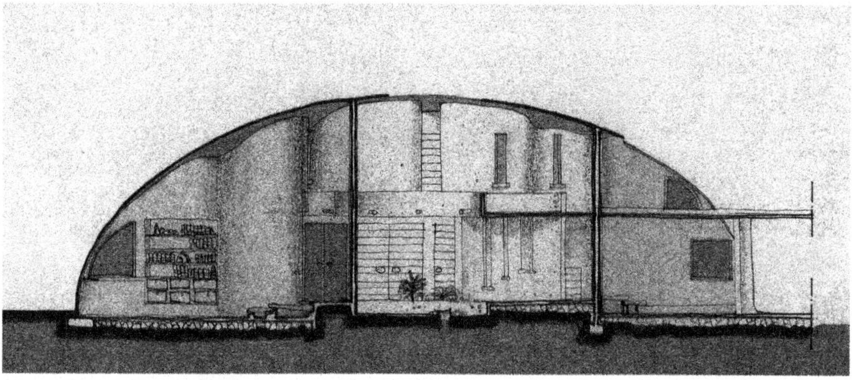

Fig. 3.15 Dante Bini, Villa Antonioni Binishell, Costa Paradiso di Gallura, 1969–1971. Section A-A (DB)

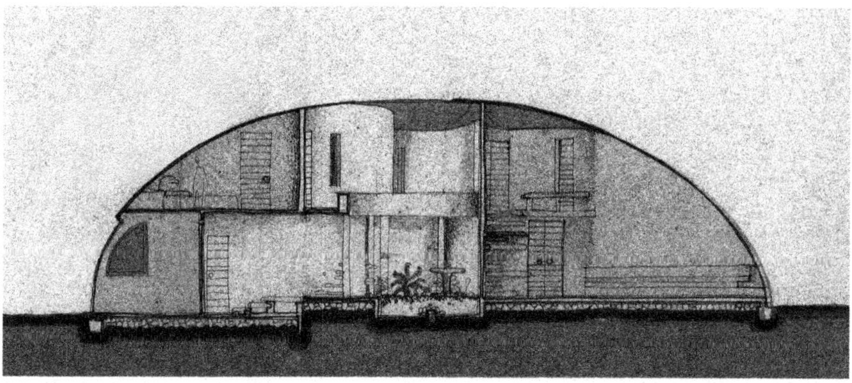

Fig. 3.16 Dante Bini, Villa Antonioni Binishell, Costa Paradiso di Gallura, 1969–1971. Section B-B (DB)

being recognised by all mainly as an inventor of new construction techniques rather than an architect.

Regarding the first design feature, the interior spaces of Villa Antonioni are mainly open towards the outside. The functions are arranged around a central core, and the volume of the cupola is subdivided into a number of rooms. The first impression is that the Binishell form is not explicitly celebrated by the design of the interiors since it cannot be experienced as a single space. However, everything becomes clearer as one moves outdoors and observes the structure from the road—that is, from above looking down towards the sea. Villa Antonioni was not designed with an introspective approach: here, the celebration of the cupola takes place at a larger scale through the contrast between the purity of the geometry of the dome and the wilderness of the surrounding landscape.

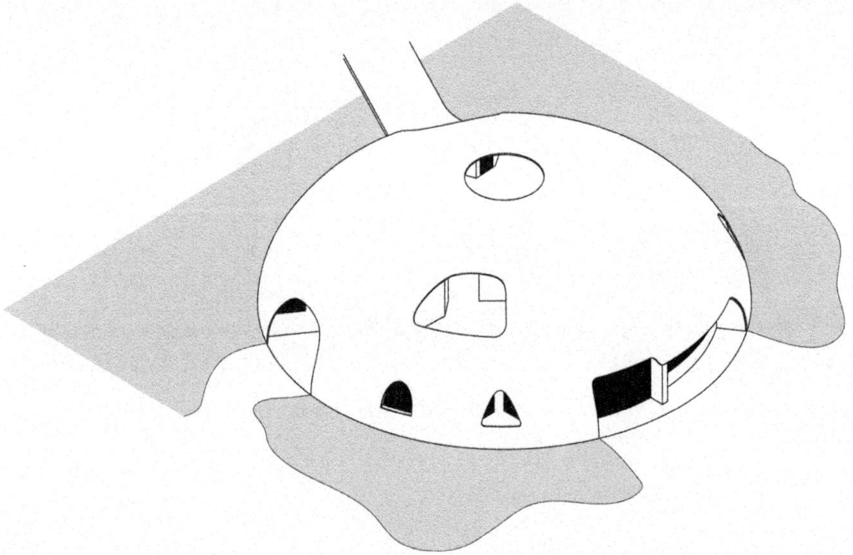

Fig. 3.17 Dante Bini, Villa Antonioni Binishell, Costa Paradiso di Gallura, 1969–1971. Axonometric view (AP)

The second design feature looks at the role of the central area of the Binishell in the overall composition of the house. Bini designed an internal courtyard illuminated by an oculus as the fulcrum of Villa Antonioni, thus positioning the circulation spaces and a small open garden in the centre of the plan. Although it seems that this solution does not fully exploit the potential of such a large void to become the main architectural feature of the house, a fair conclusion cannot be drawn without considering the project within its context. It is important to remember that Villa Antonioni is basically a device to contemplate the landscape, and the organisation of this central space further reinforces this message. The connection between the house and the island it sits on is always present, even in the most protected part of the building. Therefore, the role of this central space is to reiterate the concept that Villa Antonioni is a frame of a beautiful picture: Sardinia.

The third design feature concerns if and how the Binishell construction technique is revealed or even expressed as a tectonic feature of the house. The structural form of the dome is clearly legible in Villa Antonioni, but there are no clues to suggest a pneumatic formwork was used to erect the concrete shell. Perhaps the only elements that indicate that a modular construction system was employed here are the actual geometry of the house and the way the building sits on a sloping site. With the Binishell system, it does not matter if the project is located on a flat piece of land or on a hill. This construction method generates the same form in both cases, which explains why, at first sight, Villa Antonioni basically looks like any other Binishell structure. Although Bini cut the house openings with specific views in mind and added several elements to integrate this unique Binishell into the landscape, the visual contrast between the Sardinian landscape and the concrete shell form seems a deliberate architectural choice, which could be interpreted as another strategy to celebrate the cupola and its geometric perfection.

3.3 An Italian Design Culture of Circular Housing

Thanks to Bini's invention, the theme of the circular house has thus once again become a research topic for Italian designers, like the episodes of Mario Cavallè's igloo houses in the La Maggiolina neighbourhood of Milan, built between 1909 and 1912 with brick vaults, Pier Luigi Nervi's unrealised design widely diffused outside Italy for a circular house in 1946, prefabricated with ferrocement slabs (Fig. 3.18), or the multi-storey building for nine flats in Via Gavirate 27, again in Milan, designed in 1956 by Bruno Morassutti and Angelo Mangiarotti—whose structural design was by Aldo Favini—and built between 1959 and 1961, which is known as the "Casa a tre cilindri"—literally: three-cylinder house (Fig. 3.19).

Cavallè's igloos, measuring about 50 square metres and located on two levels (basement and first floor) within a 7.5-m diameter floor plan, were houses built for the lower and middle classes. Nervi's prefabricated house prototype, which was designed for those people who had been displaced during the war, was 11.3 m in diameter, while Morassutti and Mangiarotti worked on luxury flats measuring about 100 m^2, distributed inside circular 12-m diameter plants, that is, with similar dimensions to

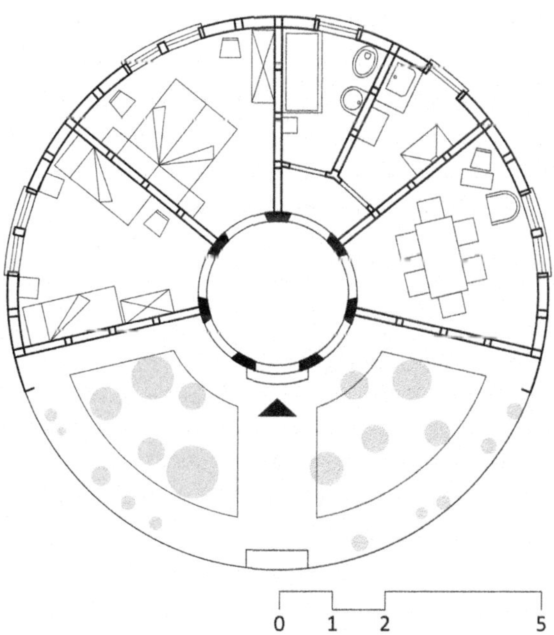

Fig. 3.18 Pier Luigi Nervi, circular house project, 1946. Plan (AP)

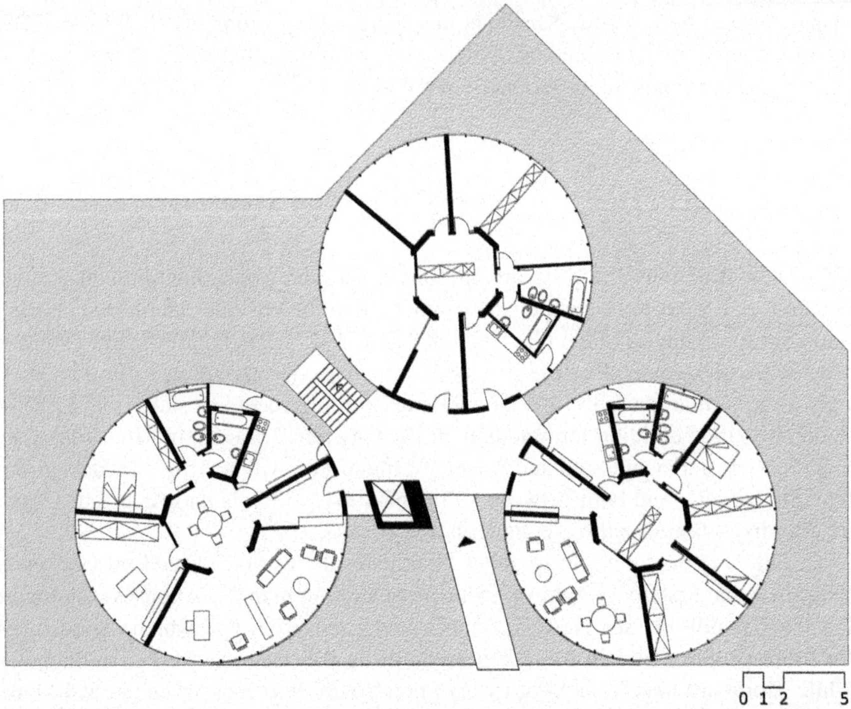

Fig. 3.19 Bruno Morassutti and Angelo Mangiarotti, Casa a tre cilindri (three-cylinder house), Milan, 1956. Plan (AP)

those used by Bini, who in fact concentrated his design research on 12-m diameter domes. As these examples indicate, the different sizes and types of users show that the circular plan can be expressed over a wide range of very different layout solutions. It is also thanking this freedom of layout, which is not only characteristic of domestic architecture, that the Binishell system is able to enjoy the commercial success it has had over the past fifty years and across several continents. "In this way, the rooms are no longer clustered and monotonous boxes, even though they are ordered together, but are now presented in a new formal expressiveness of differentiated spaces that intersect in a light and elegant spatial ductility", reports the May 1967 *Binishells spa* promotional catalogue entitled *Binishells. A new technique for new forms.* A 1969–1970*Binishells spa* catalogue also presented different housing solutions that could be placed within 12-m diameter domes, developed by the Edilizia Mediterranea firm of Naples—an example of these solutions is a holiday house project in Naples, where the various rooms are distributed around a single living-dining room with direct access from outside. Even though the spatial arrangement of the functions seems to be consistent with the Binishell form, the living room suffers from limited natural light and views (Fig. 3.20).

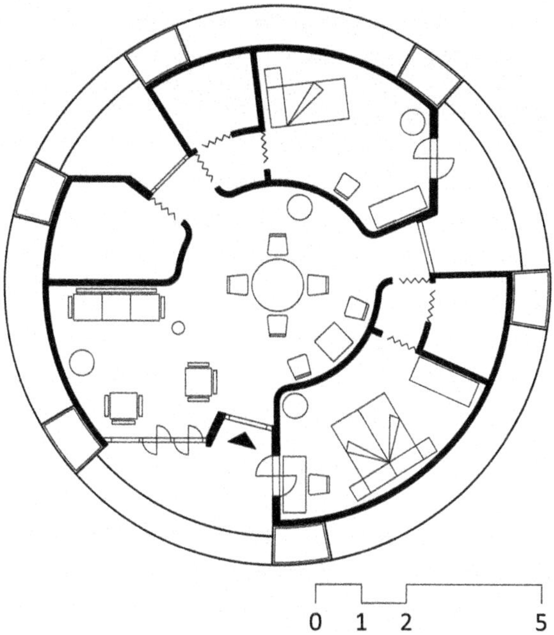

Fig. 3.20 Edilizia Mediterranea firm, holiday house project, Naples. Plan (AP)

As can be seen for the 20-m experimental house project, Bini collaborated with other fellow architects to develop a wide range of sample design applications for his patent. He aimed to promote the Binishell system as much and as widely as possible to attract private and public clients or building companies interested in licencing this innovative technique for national and international projects.

It can also be seen, from a 1969 to 1970*Binishells spa* promotional catalogue, that many of the advertised residential projects were credited to other architects, not Bini himself. This, unfortunately, prevents us from analysing such projects as the creative product of a single mind. It also becomes impossible to find points of continuity between one design proposal and another since they were developed independently by different people. Still, despite these limitations, it is important to compare a variety of residential applications of the Binishell patent because such a comparison serves to highlight the complexities of adapting the shape of a concrete dome to the scale and requirements of a dwelling project. Designing a residential Binishell is probably the most challenging task for an architect who aims to match—using Zorgno's words—"the architectural and volumetric matrix" of a building with its "constructive and structural matrix". The projects included in the 1970 promotional catalogue demonstrate, to a certain extent, how difficult it is to adapt Bini's invention to the rules, even the simplest ones, used to design domestic spaces. For example, the lack of orthogonal lines, which is not limited to the plan but extends to the elevations, seems to be an incredibly challenging hurdle to overcome.

This background is essential to understand how and why Bini developed the Minishell system for residential applications as an evolution of the Binishell. It is also reasonable to assume that this set of domestic spaces from the early *Binishells spa* promotional catalogues contributed to highlighting the limitations of the original patent and the need to revisit the shape of the dome to improve its suitability for this particular building typology.

Two of the most architecturally sophisticated design proposals in the 1970 promotional catalogue were designed by Nordio. The first design option has a half-moon-shaped layout and provides a possible solution to the issues of the holiday house project designed by the Edilizia Mediterranea firm. The living room is still located in the centre of the plan, while the other functions are compacted along half of the Binishell perimeter. This solution allowed the architect to create a large entrance opening with a terrace and provide more natural light to the interiors (Fig. 3.21). The second design proposal is for a Bungalow Binishell, capable of accommodating four double rooms under a single dome, with the toilets grouped together in the centre (Fig. 3.22). Both designs are far simpler than that of Villa Antonioni or the 20-m experimental house discussed above. Still, they show that Nordio attempted to acknowledge the presence of a roof structure that was not merely a flat slab. On the other hand, no elements of these two projects recall the employed construction technique, an aspect that Bini clearly considered in the design of his first cupola in Crespellano.

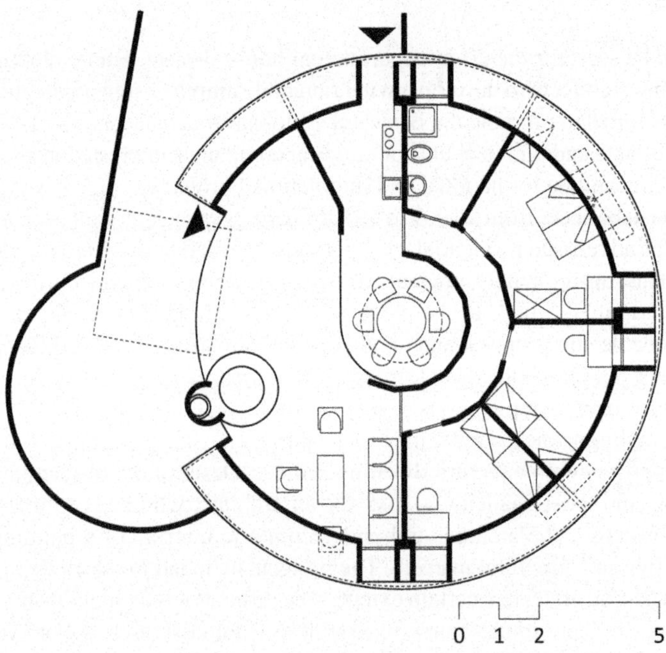

Fig. 3.21 Furio Nordio, single-family house project, 1970. Plan (AP)

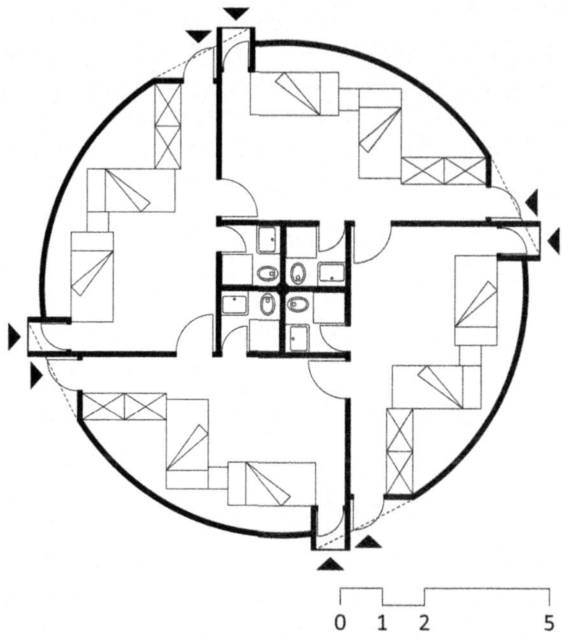

Fig. 3.22 Furio Nordio, bungalow house project, 1970. Plan (AP)

Before discussing other projects from the 1970 *Binishells spa* promotional catalogue, comparing the two design solutions developed by Nordio with the three-cylinder house designed by Mangiarotti and Morassutti will help highlight the challenges of matching the programmatic requirements of a dwelling with the structural form of a concrete dome.

The three-cylinder house was built in Milan in 1959 and takes its name from the number and shape of the building volumes that it is composed of. It consists of three identical tower elements, each containing three apartments—one apartment per floor. Since the vertical circulation spaces are incorporated in an external glazed volume that connects the three cylinders, all the apartments can take advantage of the entire floor area of the circular plan.

Like Nordio's projects, the internal spaces are organised radially around a central dining area. However, the formal similarities between the plans of Nordio's Binishells and the apartment layouts of the three-cylinder house should not mislead us into thinking that these projects feature interiors with analogous spatial qualities. A quick look at the sections clarifies that Mangiarotti and Morassutti did not have to deal with a curved ceiling profile but just simple horizontal lines. Moreover, they did not employ a dome structure, which would have transferred the forces to the ground along its perimeter. They opted for cantilevering slabs supported by a central structural core—a large column at the ground level that is subdivided into a number of load-bearing walls from the first floor up. This solution allowed the architects to use 360° strip windows, thus providing views and natural light to the apartment interiors using the entire surface of the building façade.

Even by focusing only on this single aspect of natural lighting, it is evident that a structural frame system—regardless of its design sophistication—provides much more flexibility to an architect than a Binishell. As mentioned at the beginning of the chapter, it is perfectly reasonable to design dwellings with curves, but it becomes incredibly challenging to have curves in both the plan and section. This is most likely the crucial issue that led Bini to develop his Minishell system for residential applications.

3.4 Squaring the Circle

Among the residential projects featured in the 1970 *Binishells spa* promotional catalogue, the small single-family holiday house designed by the Parisian architect Alexandre Jeleff is worth mentioning. The aspect that differentiates this house from the other design proposals in the catalogue is a square-shaped plan inscribed in the circular footprint of the Binishell; therefore, the various rooms occupy a smaller area than the surface covered by the cupola and are enclosed by infill walls. This choice allowed Jeleff to visually separate the shell structure from the orthogonal volume of the house and to rationalise the arrangement of the internal functions without compromising the integrity of the concrete roof structure. For example, doors and windows, which are standard building components, are integrated into the external walls of the dwelling and are therefore independent of the Binishell structure. Four large arch openings on the dome allow the square plan of the house to merge seamlessly with the circularity of the concrete shell, thus also preserving the overall symmetry of the composition (Fig. 3.23).

In the 1970 *Binishells spa* promotional catalogue, it is possible to see several different attempts to "square the circle", even though not all the design proposals can be considered as successful and sophisticated as this one developed by Jeleff. For example, the house project designed by the firm Bonfiglioli, Evangelisti and Vacchi completely disregards the formal features of a dome structure (Fig. 3.24). The architects used orthogonal walls and suspended ceilings to cut very functional domestic spaces from the Binishell volume and ended up designing two options that look like any other conventional one-storey house: the kitchen was placed in a barycentric position in a 12-m diameter dome, and any potential spatial complexity was avoided by inserting suspended ceilings, which meant that almost all the rooms had a flat intrados: thus, "monotonous boxes" were created in contrast with Binishells' advertised "new formal expressiveness".

Luigi C. Olivieri's Casa Cupoletta, developed for an Italian weekly women's magazine, had the aim of making the Binishell system more appealing to private clients. In this case, the architect used a cosmetic approach on the external surface of the Binishell, that is, he made up the brutal concrete cupola with colours, a chimney, a pergola and other accessories so that it looked more like to the stereotype of the single-family home (Fig. 2.16). How Olivieri ensured the magazine readers could recognise themselves with this house design is discussed in more detail in Chaps. 2

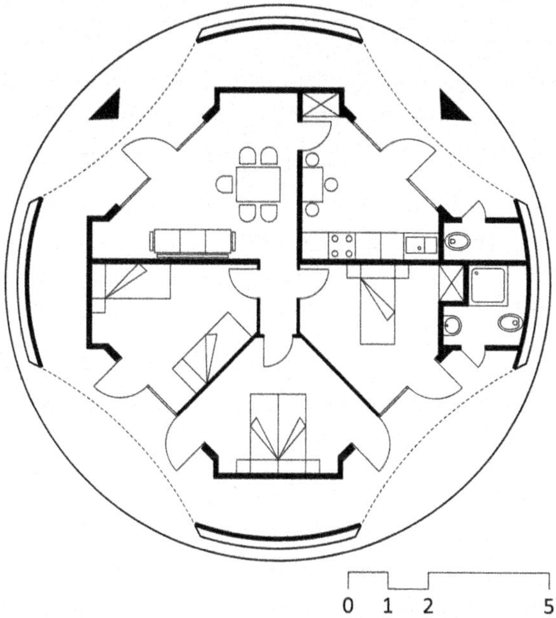

Fig. 3.23 Alexandre Jeleff, holiday house project, Guadeloupe, 1969. Plan (AP)

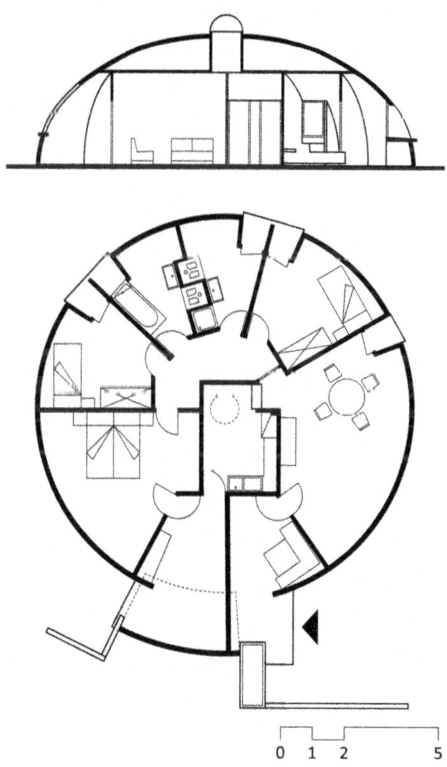

Fig. 3.24 Bonfiglioli, Evangelisti and Vacchi firm, single-family house project, Bologna, 1969. Plan and section (AP)

and 5. Here, the focus is on the plan and its relationship with the Binishell form. Olivieri was undoubtedly aware of the intrinsic difficulties of designing a house with a circular plan. Therefore, for the Casa Cupoletta, he decided to cut portions of the Binishell structure with vertical planes to work with a different footprint. The result was a cross-shaped plan, which above all allowed him to use orthogonal walls and create more entry points for natural light (Fig. 3.25). Although this is a reasonably good design option, from the functional perspective, it is hard to say whether it is a convincing piece of architecture. Once again, the architect missed what Zorgno described as the "confluence of the architectural and volumetric matrix with the constructive and structural matrix" of the building. However, it should not be surprising that a kit-home concept was paired with the Binishell system, i.e. an automated and modular construction technique developed to reduce costs and waste.

It is also not so surprising that a very similar kit-home design was used to build 40 identical single-family Binishell houses at the Torre Cintola holiday resort in Monopoli, near Bari, in the 1970s. As illustrated in two sample plans for these residences (Fig. 3.26), which were based on a 15-m dome structure, each building accommodates four small apartments of 20 m^2. Like Casa Cupoletta, these houses

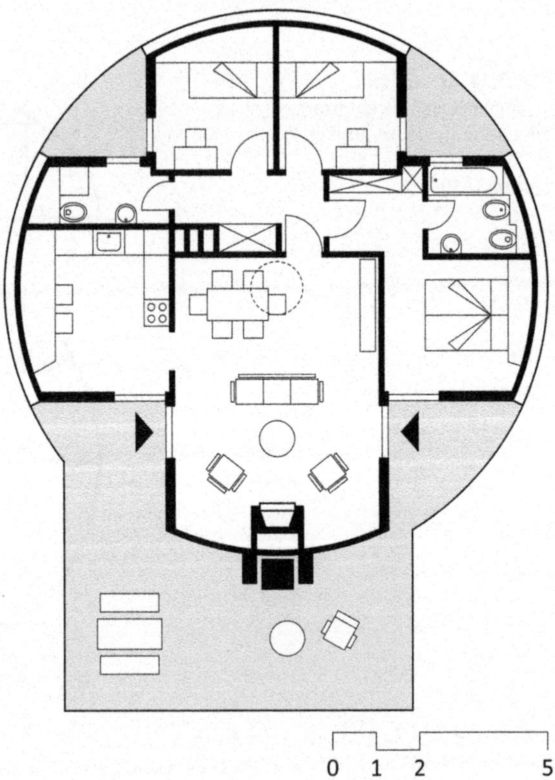

Fig. 3.25 Luigi C. Olivieri, Casa Cupoletta, 1975. Plan (AP)

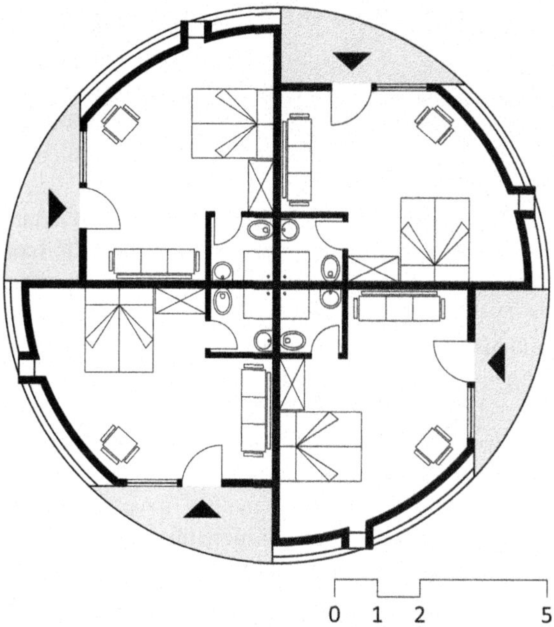

Fig. 3.26 *Binishells spa*, Torre Cintola holiday resort, Monopoli, 1970s. Apartment plan (AP)

feature vertical cuts on the dome structure and the use of orthogonal partition walls. Doors and windows were also designed using standard components.

The Torre Cintola Binishells can visually be compared to Casa Cupoletta because of the shape and position of the cuts on the dome structure. This similarity between the two projects is unlikely to be coincidental and is probably related to functional reasons. By subtracting parts of the concrete shell using vertical planes, Bini solved at least three problems simultaneously: providing access to the dwellings, simplifying the arrangement of the internal functions and creating orthogonal vertical surfaces where he could place windows and, thus, increase the natural light.

In short, Casa Cupoletta and the Torre Cintola holiday resort are projects that adapted the Binishell system to standard types of single-family houses and apartments by adding orthogonal elements to the concrete structure. They can be included among those early attempts to square the circle, which are here considered precursors of other pneumatic construction methods that Bini developed in the 1970s and the 1980s, particularly the Minishell—which was conceived and prototyped in Australia—and the Pack-Home system.

3.5 The Path to the Minishell and the Pack-Home Systems

As mentioned in Chap. 1, in 1971, the Australian minister Leon Ashton Punch was looking for a rapid system to build multipurpose centres, schools and libraries to fulfil the promises he had made during the electoral campaign. Davis Hughes, who was the general agent of New South Wales at that time and who was based in London, began investigating the matter and eventually contacted Bini. A trip to Italy was organised to visit some of the many Binishells built throughout the country, and, finally, Bini's system was considered by the Australian Government's architect of the Department of Public Works of New South Wales, Ian Thomson, to be the most suitable solution to satisfy their needs. Bini emigrated to Australia in December 1973 with his family. The initial idea was to stay "upside down" for one year and build ten concrete domes with the Binishell system, which were to be used as school facilities. Bini was appointed as an "architectural consultant to design, direct and train staff of the Department in all aspects of constructing the domes" (Pilz 1978).

We now know that Bini remained in Australia for much longer—about six years—and was able to complete far more than the initially planned ten Binishells. This was an intense period and a highlight in Bini's career, even though it was not without setbacks, which are well documented in his book *Building with Air* (2014) and in the article "Dante Bini's 'New Architectural Formulae': Construction, Collapse and Demolition of Binishells in Australia 1974–2015" (Pugnale and Bologna 2015). In summary, Bini designed five 18-m diameter domes for the Department of Public Works—three for the Narrabeen North Public School, one for the Killarney Heights Primary School and one for the Ashbury Public School—and at least twelve larger 36-m diameter domes for other schools around New South Wales. While the smaller Binishells had a variety of functions, ranging from reading areas to office spaces, the larger ones were all conceived to be used as multipurpose centres. For these 36-m domes, Bini developed two rather different design schemes (Figs. 3.27 and 3.28). The first seems particularly relevant in this context because it did not feature the classical circular plan of most Binishell projects. The structure footprint was inscribed in a square, which allowed Bini to create extra space for the storage and changing rooms outside the dome volume. Volumetrically, the Binishell is incorporated in an unfinished pyramid. On the one hand, this design proposal can be considered another attempt to square the circle. However, it is more likely that the addition of orthogonal elements to the composition was intended more to highlight and celebrate the curved geometry of the cupola.

Bini developed several new construction systems after completing his experience with the Department of Public Works of New South Wales and building several school facilities in the state. Each of these inventions responded to a specific issue that had been observed or experienced when erecting Binishells of different types and sizes in a number of countries. For example, the Minishell was developed in 1979 for small residential applications and was aimed at solving two main problems of the Binishell system: the cutting of the openings and the circular plan, which proved incompatible and inefficient for the design of low-cost homes.

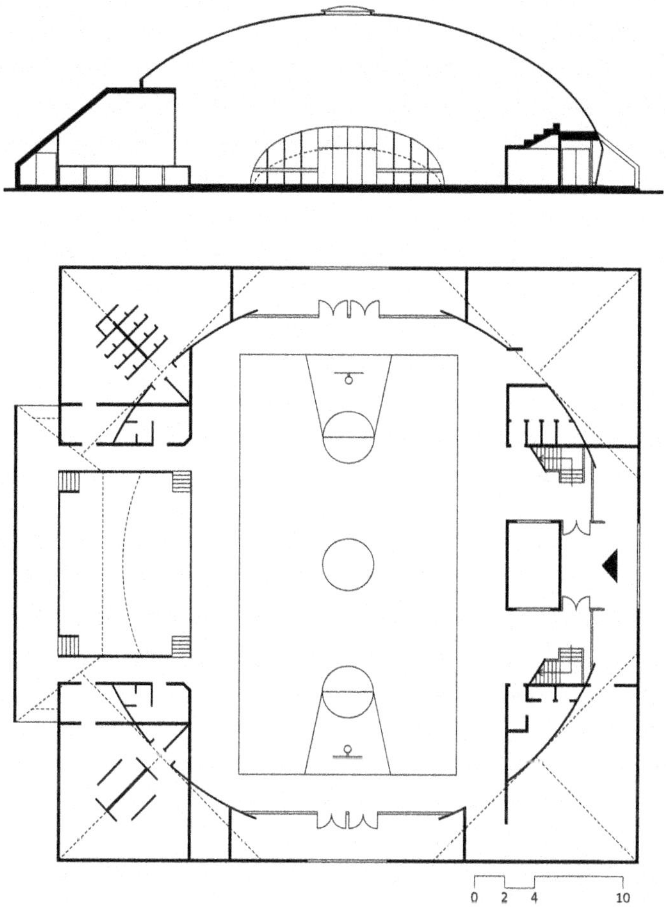

Fig. 3.27 Dante Bini, Australian 36-m Binishell project—Scheme 1, 1974. Plan and section (AP)

Five Minishells were completed in 1980 at Trinity Beach, Cairns, in Australia. These structures have been used for more than 40 years as holiday houses and are still well preserved at the time of writing. In fact, they are among the few surviving buildings in the country constructed using one of Bini's systems. Each of these five Minishells is different, but they are all divided in half to accommodate two independent apartments (Fig. 3.29).

The development of this construction method certainly stemmed from Bini's experience as a designer and builder of residential Binishells, but it was also influenced by Ross Styles' reflections on the challenges and limits of designing buildings using circular-based dome shapes. Styles graduated in architecture at The University of Sydney in 1975 and completed his thesis entitled *Binishells* entirely on Bini's work.[7]

[7] Styles was a trainee with the Government Architect's Branch of the Department of Public Works of New South Wales, between 1968 and 1973. Together with another trainee, John Rabong, he was

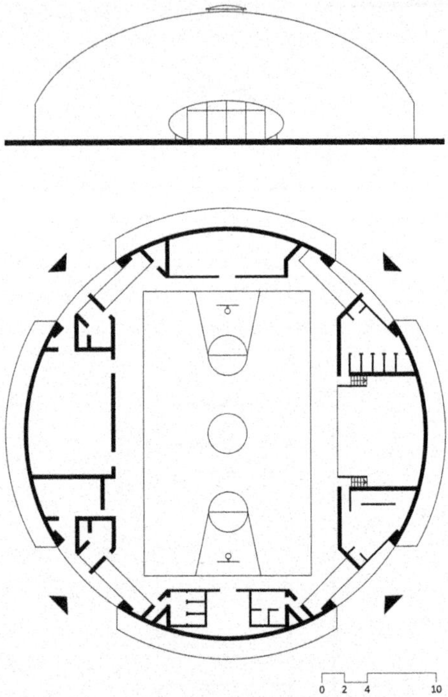

Fig. 3.28 Dante Bini, Australian 36-m Binishell project—Scheme 2, 1974. Plan and elevation (AP)

Since no one else had previously studied Bini's inventions and projects from a historical–critical and design perspective, it is not surprising that Styles' thesis offered some ideas to the Italian architect for the development of future patents.

The Minishell is basically built in the same way as a Binishell. It is a 1:1-scale form-finding and construction method in which the concrete is poured onto the ground, onto a flat surface, and is then shaped into its final form through the use of a pneumatic formwork. To develop this new system, Bini modified the shape of the inflatable membrane and its anchoring system so that the final concrete shell had a square footprint and featured four arch openings at the vertices of its quadrilateral base. The most innovative aspect was that the four openings were not cut from the final structure but were created directly during the erection process. To achieve this, Bini designed a membrane with rounded vertical edges in the shape of spinnakers to ensure that the concrete would not cover those parts of the pneumatic formwork surface during the inflation process (Figs. 3.30 and 3.31).

In a 1993 conference paper, Bini explains that the Minishell system was conceived for small projects with either an 8 by 8 or 10 by 10 square metre plan. It was a structure that could be erected in only 30 min and required 4.6 cubic metres of concrete, reinforced with 400 6 mm diameter corrugated rods combined with 36 mm diameter

asked to support to Bini and assist him with his local knowledge about construction techniques and materials for the erection of the first Australian Binishells in Narrabeen. Styles provided this information directly to the authors in 2015.

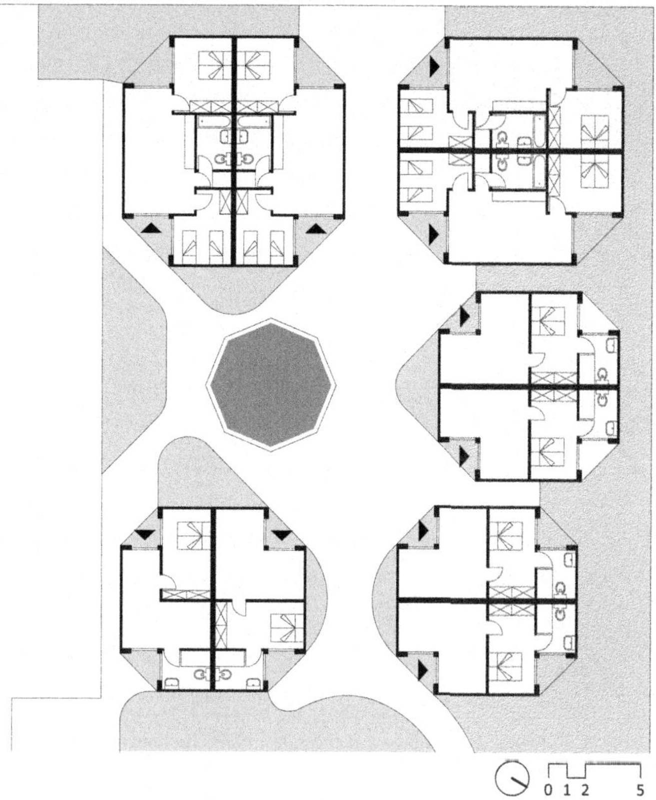

Fig. 3.29 Dante Bini, Minishell holiday houses, Trinity Beach, Cairns, Queensland, 1979–1980. Site plan (AP)

springs (Bini 1993). Even though slightly different information was published in a 1989 journal article, where Bini claimed a Minishell could either cover a 10 by 10 or a 12 by 12 square metre plan (Bini and Pietrogrande 1989), the size of this system would still be too small for applications other than residential.

Bini defined the Minishell as an extension of the Binishell concept. This is also the thesis supported in this chapter since the two techniques differ mainly in terms of dome shape rather than construction sequence and method. However, from the architectural point of view, it is possible to say that the differences between the Binishell and the Minishell extend far beyond formal matters. While the Binishell is a system that favours structural form over everything else, the development of the Minishell was directly informed by the experience Bini had gained when building residential Binishells, thus privileging practical aspects and conventions of spatial planning over geometric simplicity and monumentality.

It is possible to argue that the shape of a Minishell was derived so much from functional requirements that replacing this concrete structure with a series of conventional

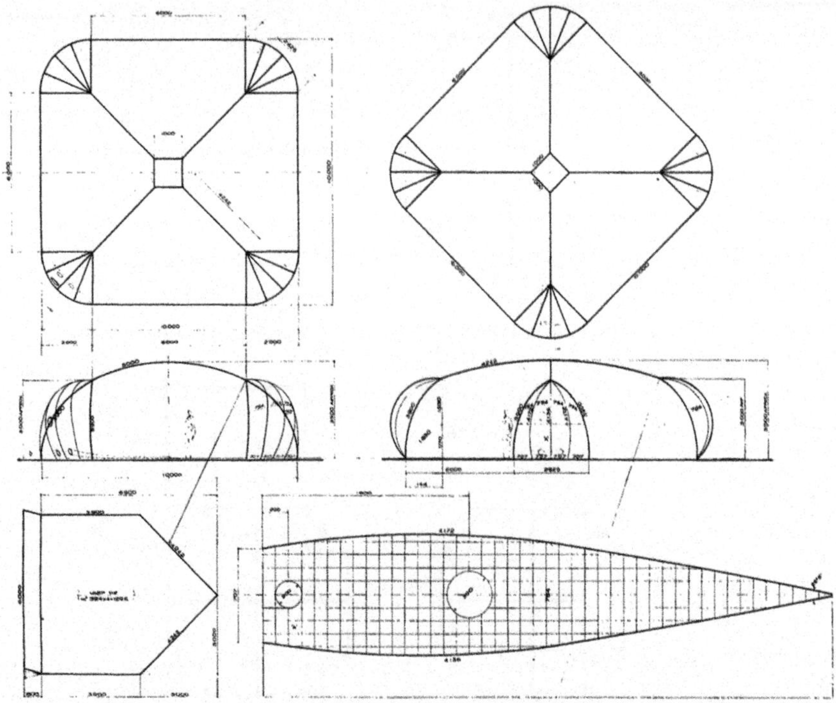

Fig. 3.30 Dante Bini, Minishell pneumatic membrane, 1979. Technical drawings (DB)

walls and a roof would not have altered the design of the interiors to any great extent. This hypothesis can be empirically demonstrated by analysing the Minishell system together with another patent that Bini developed in 1985. In that year, he was commissioned by the then Soviet Union authorities to study a pneumatic construction system for a prefabricated single-family house in order to respond to the country's urgent need for low-cost housing projects. That request led Bini to conceive a structural shelter, initially called Pack-Home and then renamed Binishelter, made of orthogonal prefabricated components, which could be mass-produced, assembled at the ground level and then erected in their final position by means of a pneumatic membrane (Bini and Pietrogrande 1989). Comparing a Minishell plan with the drawings prepared for the Pack-Home patent reveals obvious similarities between the two systems, but it also raises questions about the suitability of using shell structures to design small dwellings. At this scale, a system made of orthogonal prefabricated elements seems much more aligned with the scope of the project than a concrete shell, which needs to cover much larger spaces to be fully exploited architecturally as a structural typology (Figs. 3.32 and 3.33).

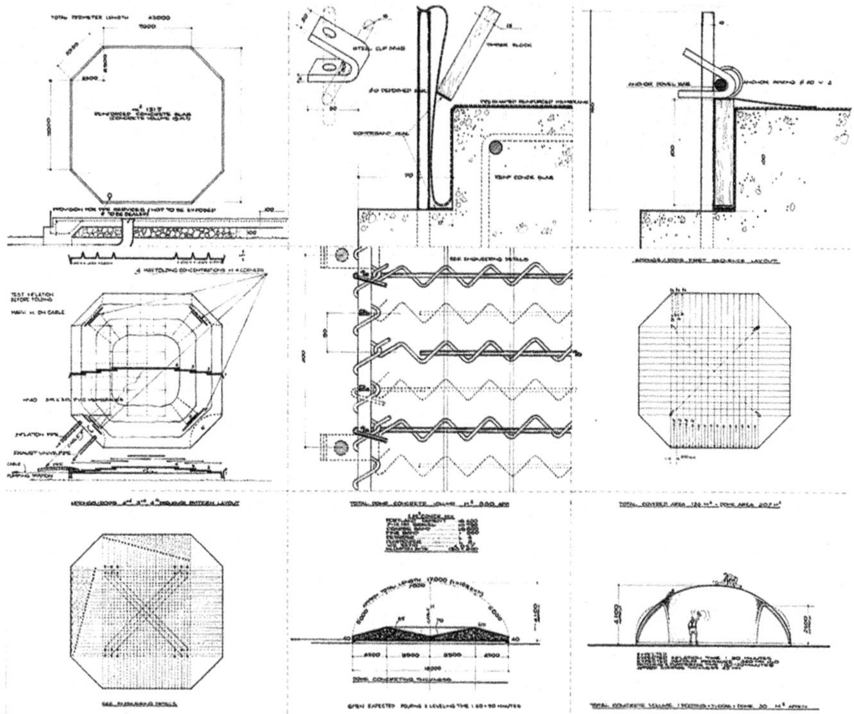

Fig. 3.31 Dante Bini, the Minishell system, November 1979. Drawings of the construction sequence (DB)

3.6 The Shape of Things to Come

Although the success of a Binishell design may depend on many factors, it is clear from the projects discussed above that some building typologies proved more flexible than others in adapting to the circular plan and curved form of the system.

An Italian *Binishells spa* promotional catalogue printed in 1967 included a blank page for potential clients to "sketch their project" and "describe their problem", claiming the Binishell company could offer "a new system to build them" and give "an answer to any building project, even yours". While this might be true for some commercial, educational and industrial applications, it is safe to say that Villa Antonioni is among the few residential projects that managed to match the architectural and volumetric matrix of the house with the structural and constructive matrix of the concrete cupola. In this project, nothing is standard or designed to be repeated a number of times or to save on costs. It is a custom-made application of the Binishell system, even though this construction technique was conceived to be modular.

Bini developed the Minishell system while staying in Australia and after the experience he had gained through the design of single-family houses and holiday

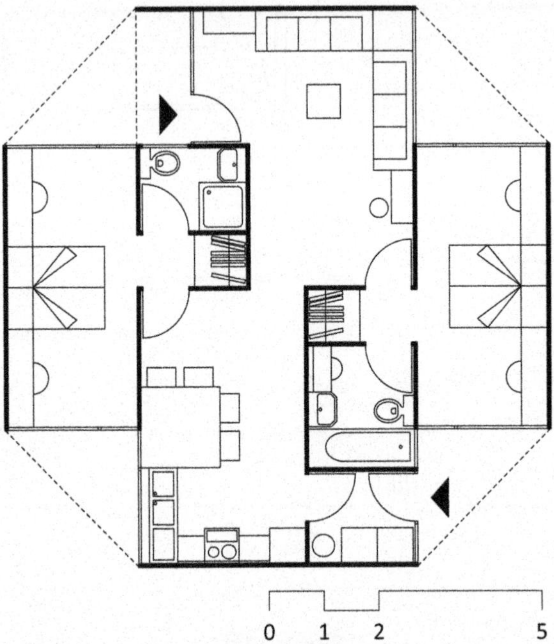

Fig. 3.32 Dante Bini, the Minishell and Pack-Home systems, 1979 and 1986. Sample plan (AP)

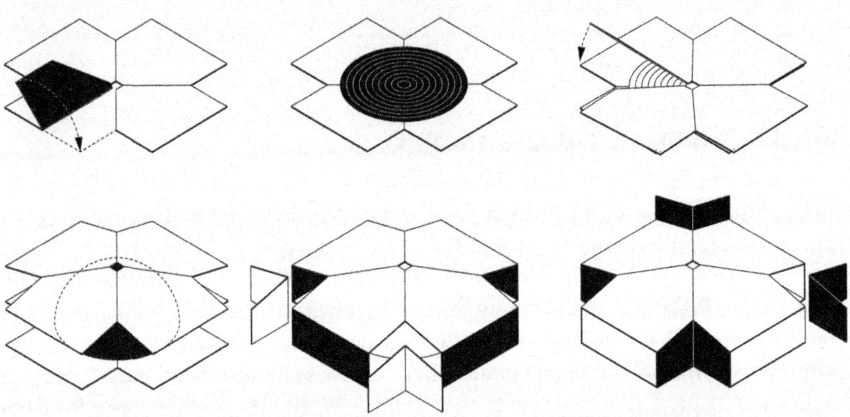

Fig. 3.33 Dante Bini, the Pack-Home system, 1986. Construction sequence (AP)

resorts in Italy. Overall, the Minishell can be considered a successful evolution of the Binishell system since it solved many of its functional issues for residential applications. However, the scale of this new patent raises a critical ontological question about the suitability of shell structures for the construction of small dwellings. The development of the Pack-Home system in 1985 suggests that more conventional structural typologies might have worked better if the main goal had been to save on construction costs and erect a large number of houses in a short timeframe. But it is reasonable to believe that Bini was not merely interested in efficiency and technicalities. He was—first of all—an architect. His inventions were mainly a vehicle through which he could explore new architectural and structural forms.

Like his father, Nicolò Bini favours a balance between design and construction over a hardcore engineering approach. At the time of writing, he advertises five pneumatic construction systems on the Binishells.com website.[8] Later on in the book, in Chaps. 4 and 5, some of these are discussed in relation to the design of the openings or the suitability of using such systems as modular elements. For the purpose of this chapter, it is sufficient to investigate whether or not Nicolò Bini deals with the relationship between architecture, structure and constructive form in continuity with his father's approach. Once again, this discussion is all about Zorgno's confluence of the architectural and volumetric matrix of a building with its constructive and structural matrix.

It is possible to notice that none of Nicolò Bini's systems preserves the circular plan and, therefore, suffers from the limits of the original Binishell patent. The absence of the circle should not be surprising since these new construction methods were primarily developed for low-cost and flexible housing projects, and it is probably for this reason that a revised version of the Minishell patent is still advertised on the Binishells.com website among the possible options for building small dwellings.

But if one excludes this system, which Dante Bini originally developed in 1979, what ideas does Nicolò Bini offer as alternatives to the cubic prison? The answer can be found in System A, a construction method for concrete shell structures that can be adapted to any planimetric shape and programmatic requirement. In many ways, and by simplifying just a little, it is possible to describe the new Binishell System A as a hybrid between the original Binishell concept and the very rational and functional Minishell. Like a Binishell, System A does not feature—nor does it have to feature—any orthogonal elements. However, like a Minishell, the forms generated through System A are informed by the shape of the plan: the house perimeter defines the boundary conditions of a pneumatic form-finding process, which is used to design the structural form of the concrete shell.

A vacuum-formed model made by Nicolò Bini illustrates how easily System A can be adapted to non-conventional and articulated residential plans (Fig. 3.34). This project features an alternation of shell parts with a synclastic and anticlastic curvature to visually separate the house rooms from the circulation spaces and create a central courtyard. It almost looks like there are four Binishells of different sizes at the corners of the plan, that blend into one another smoothly. Philip Steadman published a comprehensive analysis of this kind of circular building shapes with a courtyard, defining them as *architectural doughnuts* (2015).

On the one hand, System A can be considered as the end point of a research journey on residential concrete shells that began in 1964, with the first Binishell patent, and which was paused in 1979, with the development of the Minishell system. However,

---

[8] https://binishells.com. Accessed 15 October 2022.

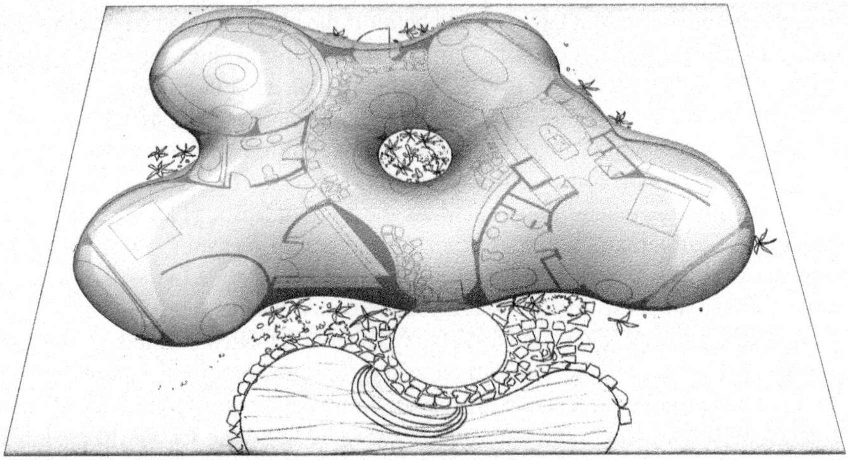

Fig. 3.34 Nicolò Bini, System-A Binishell house project, 2009. Perspective view of the digital model superimposed onto the plan drawing (AP)

on the other hand, Nicolò Bini's invention does not fully belong to the same family as the Binishell or the Minishell. System A sacrifices key features of its predecessors, such as standardisation, modularity and—as a consequence—cost-effectiveness, to provide the architect with greater flexibility to explore new structural forms that can be adapted to the needs of high-end living spaces.

Chapter 4
Open, Sesame: Cuts and Openings

4.1 The Challenges of Puncturing a Binishell

The circular plan is a characteristic of the Binishell system that, at the same time, generates design challenges and opportunities. It is not easy to arrange standard rooms, furniture and the related services within a non-orthogonal boundary, particularly in ways that are functional and efficient. However, the curved shape of a Binishell can be exploited to generate unique and custom-made spatial configurations. For example, it is possible to take advantage of the variable height of the dome or puncture the structure at specific points to create a variety of effects through natural lighting.

Circularity is a geometric feature of Binishells because a circular edge beam is required, at the ground level, to clamp the membrane used to inflate the concrete shell to its final form. As discussed in Chap. 3, this can become a limitation of designing with Binishells, particularly when the system is approached as a surrogate to conventional housing solutions. However, it is also true that any constraint can become a source of inspiration for an architect and, therefore, a design feature.

However, circularity is not the only aspect of the Binishell system that, at the same time, creates a number of challenges and opportunities. Since a Binishell is poured onto a flat surface on the ground and is then inflated into its final position, the obtained structure has no openings and, thus, is not accessible: it is an impregnable fortress. This is not a problem for conventional construction methods because they mainly involve building from the inside, under the shell. On the other hand, since the concrete structure of a Binishell is moulded from the outside, the cupola should then be perforated—literally broken—to allow access for the construction workers and users to the internal volume (Fig. 4.1).

A. Pugnale and A. Bologna, *Architecture Beyond the Cupola*,
Mathematics and the Built Environment 7,
https://doi.org/10.1007/978-3-031-26735-2_4

Fig. 4.1 Dante Bini, 30-m diameter dome at the "mushroom field", San Cesario sul Panaro, 1964–1965. Detail of the opening at the base of the shell (DB)

4.2 Form-Finding Methods and Shell Openings

Most concrete shells are designed through the use of form-finding methods that automatically create openings. For example, the reverse hanging method is carried out using a piece of fabric or a rigid net suspended from support points or edges. A structurally optimal shell is literally found by leaving the model under a gravity load and inverting the resulting geometry. With this method, catenary openings are generated in between supports (Fig. 4.2).

Heinz Isler designed several concrete shells through the reverse hanging method. The Deitingen service station on the Bern-Zurich motorway, completed in 1968, is one of his finest examples (Chilton 2000: 92–95). The project consists of two identical concrete shells with a triangular plan, which provide shelter to the refuelling area. The two structures are clearly visible from the motorway and give a sense of openness and lightness. Form finding was only performed on one structure, which

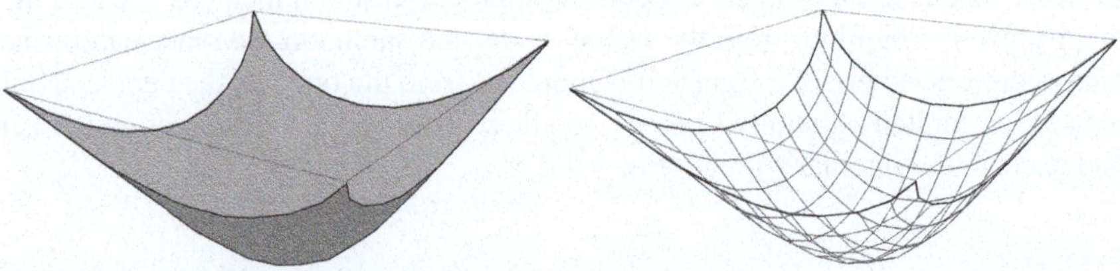

Fig. 4.2 Reverse hanging method. Form-finding schemes with fabric, on the left, and a rigid net, on the right (AP)

was then mirrored to create the second one. Both shells are supported at three points, positioned at the vertices of the triangular plan, and feature large arches in between the supports, generated during the form-finding process.

The reverse hanging method could also be used to create structural forms without openings, but this would require the physical model used for form-finding to be fully constrained along the base. This is the case with many gridshell structures that cover internal courtyards, such as the Great Court of the British Museum in London, which was completed in 2000 (Williams 2001), the Hamburg History Museum (Holgate 1997: 110–115) or the more recently completed National Maritime Museum in Amsterdam (Adriaenssens et al. 2010: 112–119) and the Robert and Arlene Kogod Courtyard at the Smithsonian American Art Museum in Washington DC (Peters 2007). Although it is less common to find structures designed through the reverse hanging method that are fully constrained along the edges, it is possible to mention a couple of well-known examples: the experimental timber gridshell designed by Frei Otto in Essen in 1962, which has a simple dome-like synclastic shape on a superelliptical plan, and the more complex 1975 gridshell of the Mannheim Multihalle (Fig. 4.3) (Liddell 2015). This latter project only partially falls into this category: the edge beams are almost flat—although they are raised off the ground—and there are a few catenary arches along the boundary to provide access to the building. However, from the design point of view, it is worth highlighting that the reverse hanging method only generates this kind of funicular shapes when a rigid net is constrained along its whole perimeter. It is more common to see designs where the structural model is suspended from a few points along the edges or the vertices, thereby creating arches as part of the form-finding process.

Designing structural forms through the use of well-known geometries, such as spheres, cylinders and hyperbolic paraboloids, can lead to similar results. Eero Saarinen's Kresge Auditorium, built at the Massachusetts Institute of Technology (MIT) between 1953 and 1955, is an example of a shell structure generated by cutting three orthogonal planes in a sphere (Fig. 4.4). This operation created three large openings that were infilled with glazed facades (Foxe 2010). From the geometric point of view, the final shape has some features that can also be found in Isler's project for the

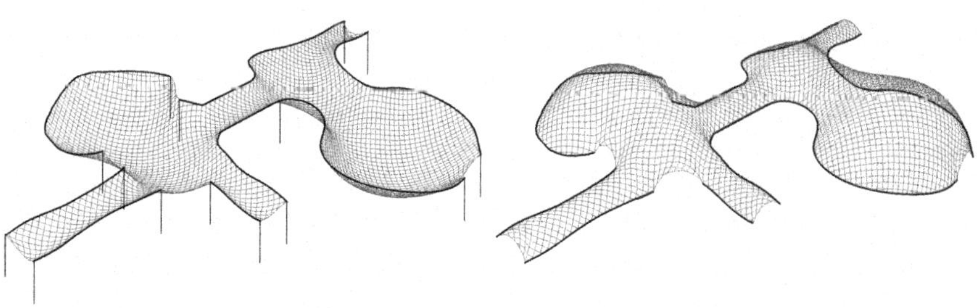

Fig. 4.3 Frei Otto, Mannheim Multihalle, 1975. Form-finding schemes (AP)

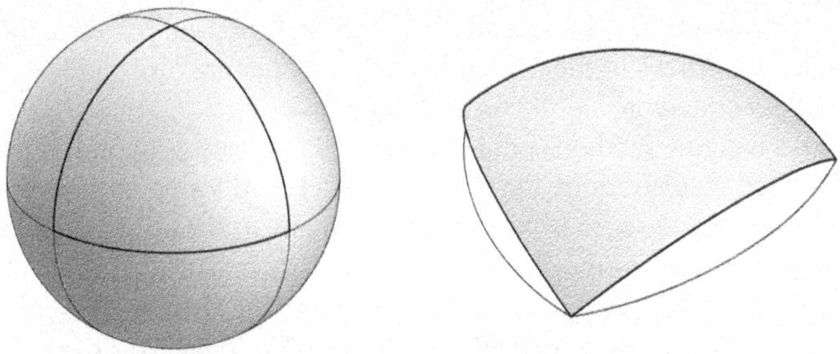

Fig. 4.4 Eero Saarinen, Kresge Auditorium, Massachusetts Institute of Technology (MIT), Cambridge, 1953–1955. Schemes of the spherical dome shape (AP)

Deitingen service station: both shells are synclastic surfaces with a triangular footprint, which are only supported at the three vertices. However, a mere formal resemblance is not enough to make the two projects analogous from the design process point of view or even in terms of architectural result. In this text, the geometric similarity is only highlighted to draw attention to possible parallelisms between different ways of designing openings in shell structures. It is important to recall that the two form-finding strategies used in these projects generally serve different purposes. The reverse hanging method generates optimal structural shapes, which, however, can be challenging to calculate and build. On the other hand, geometric forms are usually employed to simplify the design and rationalise the construction of lightweight shell structures, which was the case of the Jørn Utzon's Sydney Opera House project, where concrete sails were shaped using wedge-shaped cuts from a single sphere.

Both synclastic and anticlastic surfaces can be employed to design shells that do not require any additional structural cuts in order to be used and inhabited. Hyperbolic paraboloids, sometimes referred to as saddles, are an example of anticlastic surfaces, which became popular among architects and engineers of the 1960s. Félix Candela is highly regarded as the designer who exploited the potential of such a geometry the most in architectural and structural design. His 1958 project for the Cuernavaca Chapel, in Mexico (Fig. 4.5), is a simple and elegant application in which a single hyperbolic paraboloid was trimmed with two inclined planes to create a horn-shaped structure with two wide openings at the base, facing opposite ends (Draper et al. 2008: 128–141).

Hyperbolic paraboloids can be trimmed in two main ways, depending on the orientation of the cutting plane, and this trimming can lead to either a linear or a curved edge. Curved edges are featured in the Cuernavaca Chapel, while linear edges were chosen for the 1959 Church of San José Obrero in Monterrey (Fig. 4.6). The latter is a simple composition of two identical hypars supported at two vertices: this configuration creates two large openings at the base of the building, but also a skylight in between the hypars. Although the shells designed from geometric forms usually have openings at the base of the structure, there are several projects in which the composition of spherical surfaces or hyperbolic paraboloids also creates cuts in

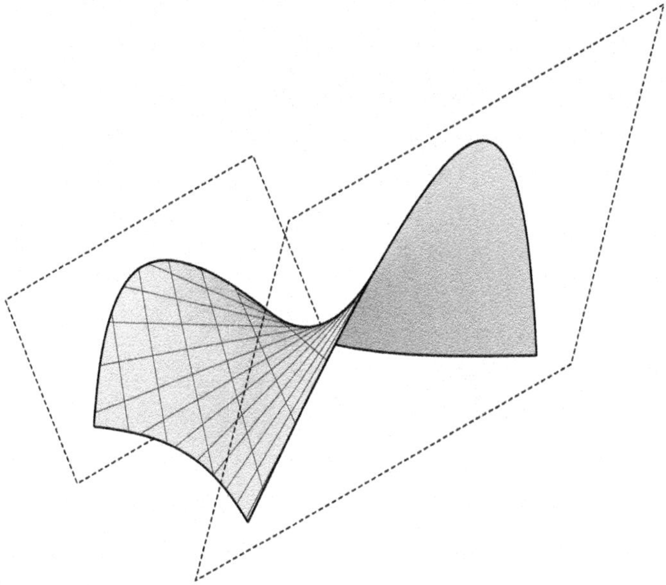

Fig. 4.5 Félix Candela, Cuernavaca Chapel, Morelos, 1958. Scheme of the hypar structure (AP)

the top part of the building, such as in the Sydney Opera House or the aforementioned Church of San José Obrero.

In the Cathedral of St. Mary of the Assumption in San Francisco, designed by Pier Luigi Nervi and Pietro Belluschi between 1963 and 1971, hyperbolic paraboloids were trimmed using vertical planes to generate modular structural elements that could be combined thanks to their orthogonal edges (Fig. 4.7). Eight hypars were paired

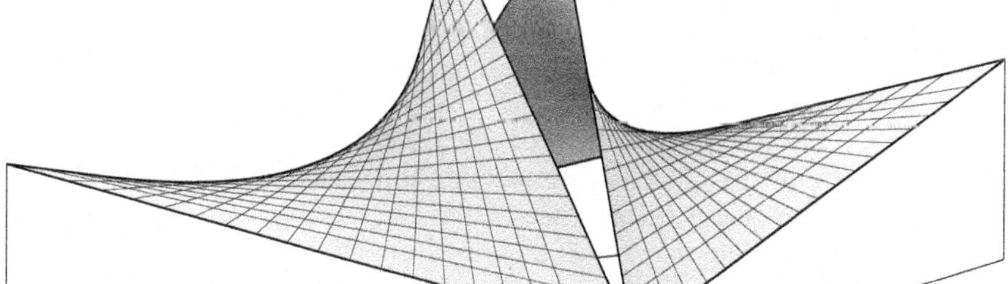

Fig. 4.6 Félix Candela, Church of San José Obrero, Monterrey, 1959. Scheme of the hypar structure (AP)

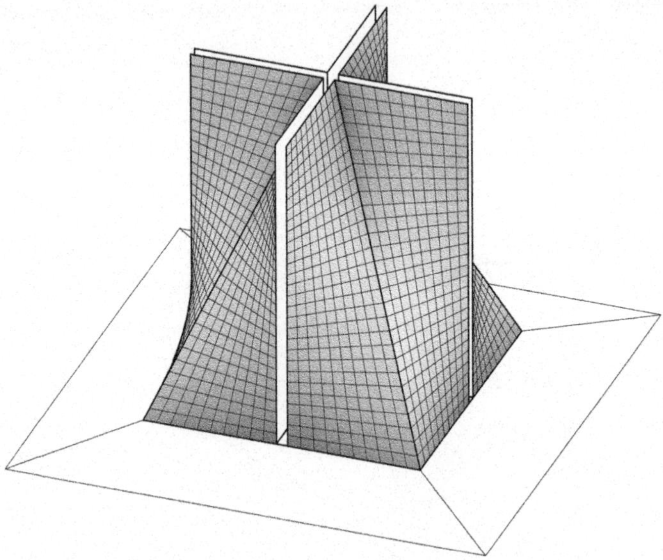

Fig. 4.7 Pier Luigi Nervi and Pietro Belluschi, Cathedral of St. Mary of the Assumption, San Francisco, California, 1963–1971. Scheme of the hypar structure (AP)

and mirrored twice to define a symmetrical layout which—in plan—symbolically draws a cross through blades of light. This is another project in which geometric forms were used to create openings that were not placed on the base, but above the space generated by the shells.

The 1958 Los Manantiales restaurant at Xochimilco in Mexico is one of Candela's most popular projects. From the geometrical point of view, it is a polar array of a single hypar module, trimmed using one inclined plane and two vertical planes. The former generates the curved profile visible at the perimeter of the structure, while the vertical planes create linear edges that allow two adjacent hypars to be merged into a single surface. The trimming and composition of the hyperbolic paraboloids define the distinctive feature of the building, namely eight arch openings at the base of the structure (Fig. 4.8).

The methods used to design tensile structures, or more in general minimal surfaces, all follow a similar logic regarding how the openings are created as an integral part of the form-finding process. It is worth recalling that minimal surfaces generated from a single closed frame, either using soap bubbles or fabric sheets, are always anticlastic saddle shapes that closely resemble the geometry of the hyperbolic paraboloid. Therefore, it is easy to draw formal parallelisms between Candela's restaurant at Xochimilco and the 1957 Dance Pavilion designed by Otto in Cologne for the Federal Garden Exhibition (Fig. 4.9) (Pugnale 2018). Although the former is a concrete shell conceived to use hypars, and the latter is a tensile structure, hence a minimal surface, both designs have wavy façades with large openings that result from two different structural-form generation methods. The circular plan is also an

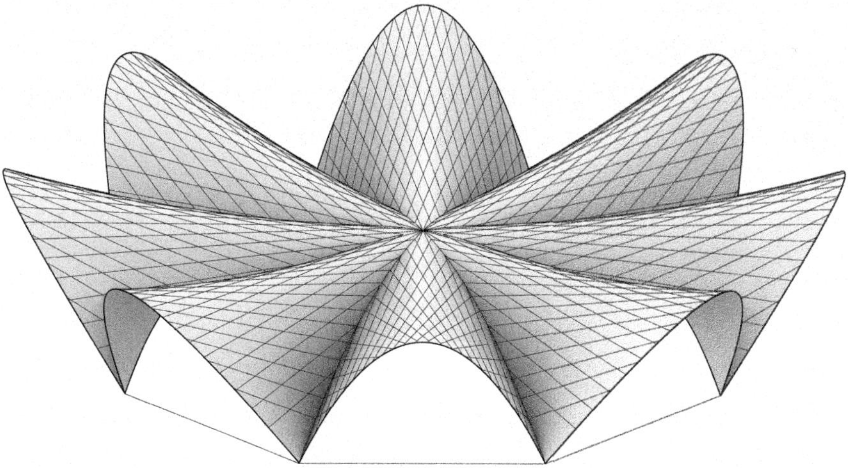

Fig. 4.8 Félix Candela, restaurant, Xochimilco, Mexico City, 1958. Scheme of the hypar structure (AP)

important design feature of both projects, thus making them even more relevant in this chapter dedicated to openings and cuts in Binishell structures.

There are also form-finding techniques that do not generate structural shapes with openings. For example, the "inflated hill" method, invented and fine-tuned by Isler, has been used to design concrete shells by inflating a rubber or latex membrane clamped to a boundary frame of the desired shape. Since the membrane must be airtight to work, the openings can only be trimmed at a later stage, after the final shell geometry has been achieved (Fig. 4.10) (Chilton 2000). It is easy to associate the Binishell system with Isler's "inflated hill" technique since air is the means that is used in both cases to achieve structural form-finding. However, it is important to

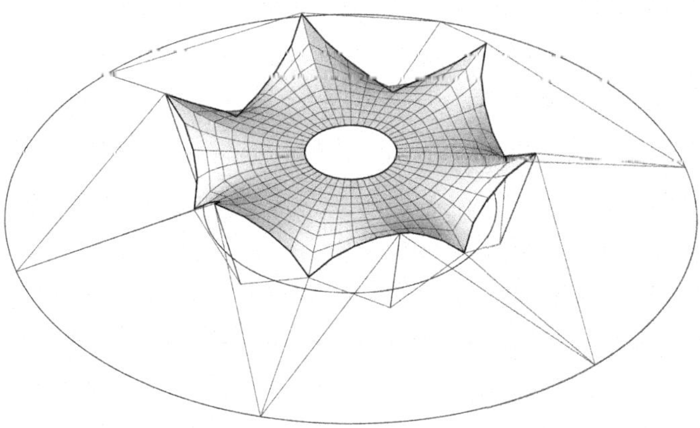

Fig. 4.9 Frei Otto, Dance Pavilion, Cologne, 1957. Scheme of the tensile structure (AP)

Fig. 4.10 Heinz Isler, Pneumatic form-finding or "inflated hill" method, developed in 1954. An original model photographed in Isler's office in Lyssach in 2003 (ER)

clarify why this analogy is conceptually incorrect. The Binishell was not conceived for scale models and should not be considered as a form-finding method. It was invented to blend structural efficiency with construction automation principles in order to build concrete shells rapidly and cost-effectively. It is a system that addresses the main issues of concrete structures designed through reverse hanging or inflation, i.e. the geometric complexity and waste generated by the formworks. However, for the purpose of this chapter, the most relevant consequence of the fusion between structural design and construction concerns the possible ways openings can be created in a shell using this system.

4.3 How to Cut a Binishell

Bini approached the design of openings in Binishells in a variety of ways. Ross Styles described these approaches in his thesis (1975). Styles indirectly wrote about openings while discussing opportunities for natural lighting. He identified two main options, that is, with natural light either entering from above the space formed by the Binishell or from along its perimeter. These two conditions also determine whether the structure is lit directly or indirectly.

4.3.1 Direct Lighting

Styles began his review of approaches to the design of natural light in dome structures by considering key historical buildings. The first one he discussed was the Pantheon in Rome, as it features an uncovered oculus at the apex of a hemispherical dome, which Bini reproposed several times in his Binishell projects.

It has never been a secret that the Pantheon inspired Bini, who even wrote about it in his books (Bini 2009, 2014). This inspiration is also attested by some of his early prototypes and his very first Binishell project in Crespellano—completed in July 1964—which featured a revisited version of the Pantheon oculus. Many other elements in this Crespellano Binishell suggest that this circular opening at the apex of the cupola was a deliberate architectural gesture. For example, the pattern created by the steel reinforcement imprint on the internal surface of the shell defines concentric rings that emphasise the shape, size and position of the oculus. The Binishell was finished externally with a circular crown that allowed the presence of this symbolic opening to be perceived, even though it was not directly visible from the outside (Fig. 4.11).

The 30-m Binishell prototype that Bini built in the "mushroom field" in 1965, to demonstrate to Mario Salvadori that his pneumatic construction system really worked, is another example of a concrete cupola that features an oculus. In this case, the reference to the Pantheon is even more obvious as the two domes have a similar scale (Fig. 4.12).

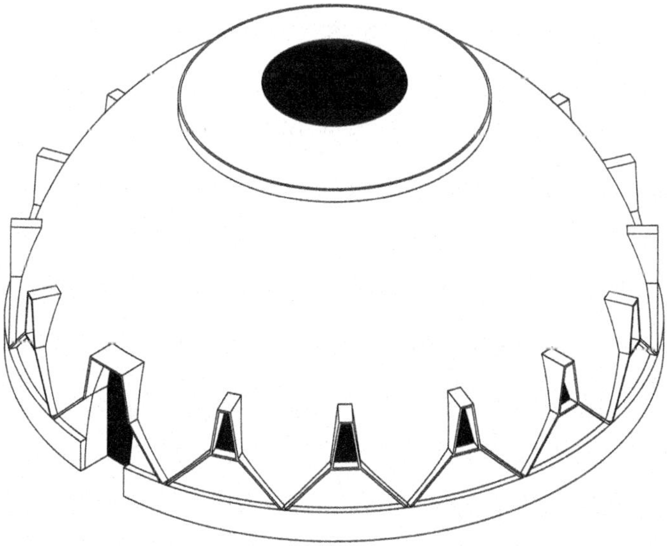

Fig. 4.11 Dante Bini, Crespellano Binishell, Unipack company, July 1964. Axonometric view (AP)

Fig. 4.12 Dante Bini, 30-m diameter dome at the "mushroom field", San Cesario sul Panaro, 1964–1965. Internal view (CD)

Villa Antonioni in Sardinia is another Binishell that is characterised by a circular opening positioned at the apex of the dome. However, in this case, the internal spaces are arranged around a central courtyard, and the oculus is only visible from the outside. It does not provide light or air directly to the house—for example, to the living room or the bedrooms. This might seem like a bizarre or difficult choice to explain from an architectural point of view. However, considering that Villa Antonioni is one of the few Binishells conceived for a specific site, which was also a project commissioned by a private client, it becomes clear that—in this project—the design of the openings was informed more by the framing of specific views of the landscape than by an idea of monumentality (Figs. 4.13 and 4.14).

However, for the more recent Binishells, it could be argued that the presence of an oculus has been a direct consequence of the improvements made to the concrete compaction technique over the years. The Binishell patent has been revisited several times, and surface vibrators were introduced at a certain point to compact the fresh concrete immediately after inflation. Such vibrators were manually controlled from the ground level using ropes that rotate around a pivot placed on the top of the shell.

It could be intriguing to create the myth that this pivot was the actual generator of the Binishell oculus and to believe that such a symbolic element was the remnant of a concrete compaction technique. However, the reality is that Bini wanted to recreate a contemporary version of the Pantheon, and this is attested by the presence of the oculus, right back from the early Binishell projects, even when concrete compaction did not involve the use of surface vibrators, pivots or ropes (Fig. 4.15).

Fig. 4.13 Dante Bini, Villa Antonioni Binishell, Costa Paradiso di Gallura, 1969–1971. View of the internal courtyard and the shell oculus (RC)

A design proposal for a large congress hall, which Bini developed in 1987 for the then Soviet Union, but which remained on paper, shows how the concept of monumentality was at the centre of many Binishell projects for public and government buildings of a certain scale. This Binishell does not feature an oculus, but this symbolic element was replaced by a spotlight projected towards the sky, and non-structural ribs that visually connect the support point to this source of light (Fig. 4.16). A perspective sketch that was curiously drawn on a darkened photo of the 1978 Sports Dome at Malvern Girls' College in Edinburgh illustrates how this project would have looked like at night. Here, monumentality can be interpreted as a celebration of the cupola, as emphasised by its geometry and scale (Fig. 4.17).

The oculus is often present in large Binishell projects, as in the sports centre designed by Marco Meozzi and Paolo Pettini in 1976–1977in Prato, near Florence. Even though Bini did not directly conceive this 32-m dome, it is reasonable to assume that the hole at the apex of the structure performs more than just the function of providing natural light to the indoor pool. This can be deduced from the arrangement of the services and from other panels that are suspended from the ceiling, which form concentric rings that emphasise the shape of the Binishell and of the opening towards the sky (Fig. 4.18).

Fig. 4.14 Dante Bini, Villa Antonioni Binishell, Costa Paradiso di Gallura, 1969–1971. View of the sea from the strip window (PA, DRS)

4.3.2 *Indirect Lighting*

However, not all the Binishells were built with an oculus. Whether or not there is an opening at the apex of the dome seems to depend on many factors, including the function of the building, the arrangements of the internal spaces and the architect's approach to the design of natural lighting.

For example, the Binishell at Pegola is one of the early prototypes that did not feature this type of opening. This cupola was cut at the base to create doors and windows that provided access and natural light to the building. Styles described this design approach, which is conceptually opposite the previous one, by referring to the dome of the Hagia Sophia Grand Mosque in Istanbul, where forty arched windows visually separate the base building from its cupola, thus allowing visitors to perceive the structure as extremely light, that is, as almost floating.

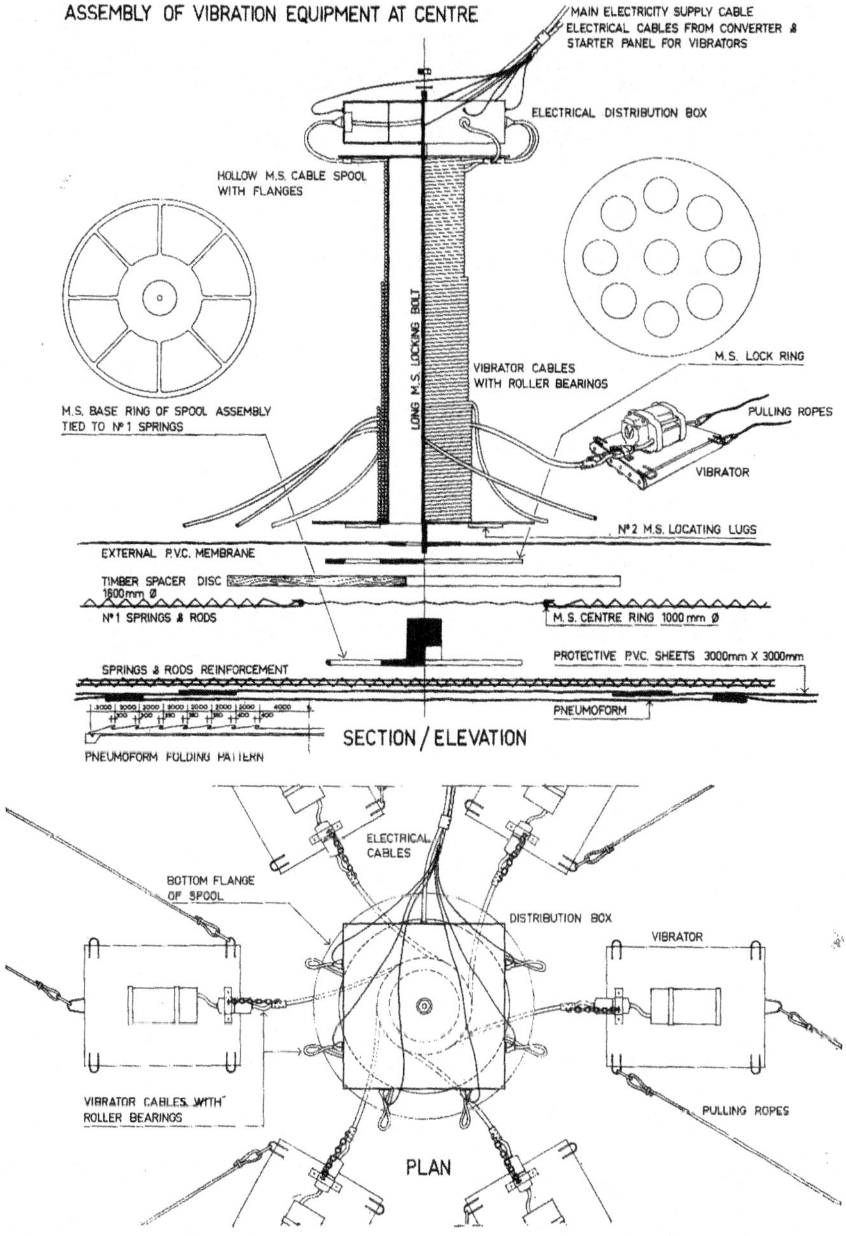

Fig. 4.15 Dante Bini, improved concrete compaction method for the Binishell system, 1974. Construction schemes (DB)

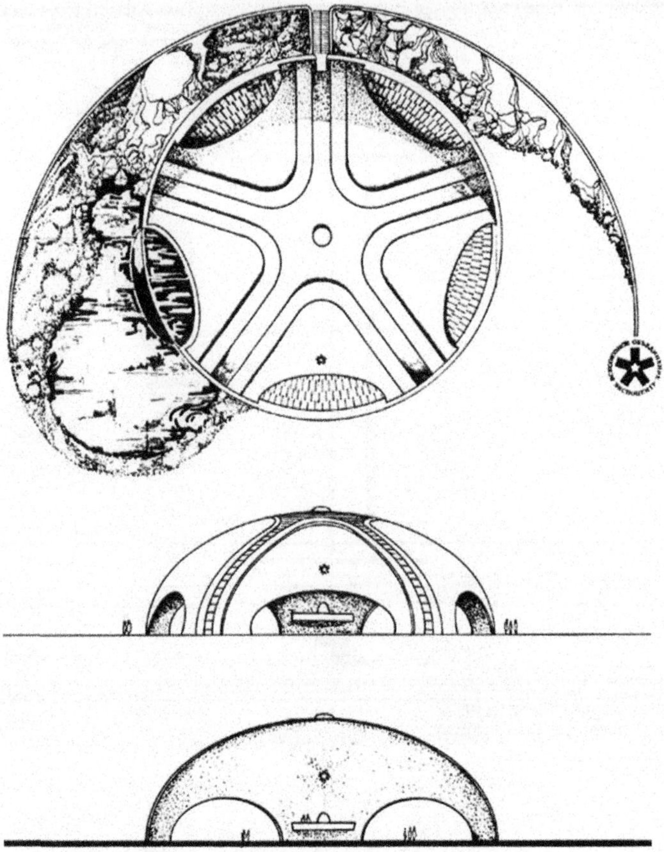

Fig. 4.16 Dante Bini, Binishell project for a large congress hall in the Union of Soviet Socialist Republics (USSR), 1987. Plan, section and elevation (DB)

The Binishell at Pegola was most likely inspired by this or another similar design precedent. From the construction point of view, it is worth mentioning that the Pegola project was one of the few concrete structures that Bini inflated using a hemispherical membrane, which created an angle between the ground floor and the shell that in turn simplified cutting the openings and inserting prefabricated structural elements to support the dome. This structure had a 12 m diameter and featured 18 openings placed along its perimeter (Fig. 4.19).

A similar design strategy can be found in a 1968 Binishell project for the Hotel Ariston Molino Buja in Abano Terme. Although this cupola is larger and not hemispherical, and it again features a series of openings along the perimeter of the structure, it was infilled with the same prefabricated arch elements. In this case, the inner surface of the shell was also decorated with the symbols of the zodiac and constellations, along with a motif of parallels and meridians that follows the rhythm of the openings and emphasises the shape of the dome (Fig. 4.20).

Fig. 4.17 Dante Bini, Binishell project for a large congress hall in the Union of Soviet Socialist Republics (USSR), 1987. Sketch on a photograph of the Binishell at Malvern Girls' College, Edinburgh (DB)

4.4 The Connection to the Ground

An important consequence of cutting the Binishell at the base, especially when this cut involves the entire perimeter of the structure, concerns the design of the connection to the ground. Since the dome is no longer in direct contact with the earth and instead sits on another structure—a pedestal that acts as a buffer—the designer is faced with a series of new opportunities to define how the dome touches and integrates with its surroundings.

Styles discussed this issue in his thesis and argued that a well-designed Binishell should not simply sit on a flat piece of land as if it were an object dropped from the sky. It is easy to imagine the origin of this statement from an architectural point of view: even classical columns are tripartite and use a base to separate the ground from the actual structural element. Therefore, the same reasoning can be applied to a Binishell, which can be regarded as a structural element that can be transformed into an architectural composition and integrated with its context and other volumes.

Styles continued by describing the design strategies that had been adopted—at least up to the time of writing his thesis—to solve the connection of a Binishell to the ground. He presented various options, ranging from submerging the dome underground to lifting the shell from the earth, but despite the apparent diversity of

Fig. 4.18 Marco Meozzi and Paolo Pettini, Binishell swimming pool, Prato, 1976–1977. Internal view (DB)

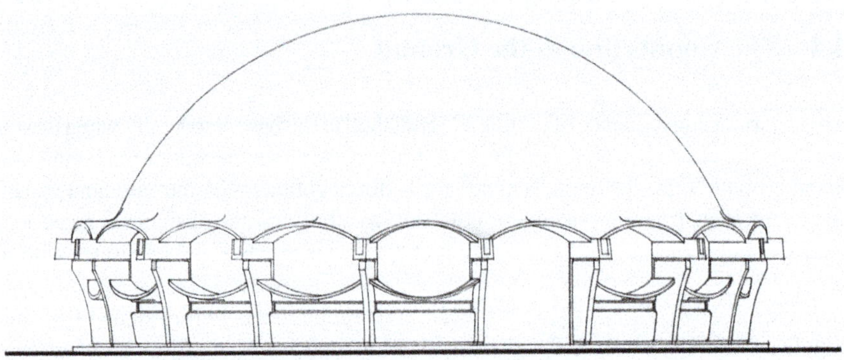

Fig. 4.19 Dante Bini, Pegola Binishell, 1964–1965. Section (DB)

such options, they all fall into two main categories: concealing the connection to the ground or removing the physical contact between the concrete and earth by adding other building elements.

When discussing the connection to the ground in relationship with Binishell openings, it is particularly important to highlight how aspects such as mass and lightness

Fig. 4.20 Dante Bini and the Culligan construction company, swimming pool Binishell for the Hotel Ariston Molino Buja, Abano Terme, 1968. Internal view of the cupola (HAMB)

have been addressed by Bini—and other architects—along with the issues concerning the design of natural light and view framing, as already mentioned above.

4.4.1 Concealing the Connection to the Ground

Examples of Binishells in which the dome is partially covered at the base by other building parts can be found in some Australian projects. As already mentioned, Bini emigrated to Australia in the 1970s with the intention of building a number of multipurpose school facilities quickly. The structures he constructed there were all based on two design schemes, which he called "Scheme One" and "Scheme Two". The Narrabeen North Public School, which features three 18-m Binishells, is the only exception, as all the others had 36-m domes.

In the Scheme One plan, the cupola was inscribed within a square and covered a multipurpose area that included a basketball court and a stage. Four annexes at the corners of the square were dedicated to storage spaces and changing rooms. Access to the facility was via three large arch openings, each located on a side of the building, halfway between one annexe and the other.

In this symmetrical composition of volumes, the three entrances and the top of the dome were the only parts of the Binishell that were visible from the outside. The remaining elements of the connection to the ground were concealed by the volumes surrounding the structure. These four volumes, positioned at the corners of the square plan, had pitched roofs, the shape of which recalled that of an unfinished pyramid that incorporated the Binishell. The faces of this hypothetical pyramid were tangent to the surface of the dome, which defined their inclination and angle from the ground plane (Figs. 4.21 and 4.22).

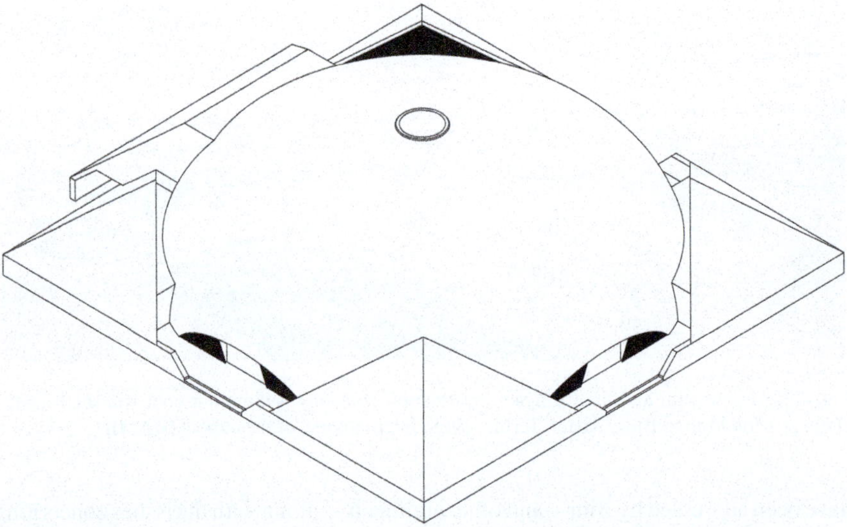

Fig. 4.21 Dante Bini, Australian 36-m Binishell project—Scheme 1, 1974. Axonometric view (AP)

Fig. 4.22 Dante Bini, Georges River College multipurpose Binishell, Peakhurst, New South Wales, 1974. Photograph of the building a few days before demolition, 12 May 2014 (AP)

From the functional point of view, this was a successful design scheme because it allowed those parts of the shell that had a low internal height to be used. From an architectural point of view, this design addressed the connection to the ground by concealing most of the base of the Binishell, except for the three entrance doors. It was a design response which—once again—enhanced monumentality and can be considered as a celebration of the cupola.

The idea of surrounding the Binishell with other simple geometry volumes can already be seen in one of the very early concepts that Bini published in an undated catalogue to promote his construction system. Bini does not recall whether this project was conceived for a museum or another public building, but he has undoubtedly confirmed this was one of his first design proposals featuring a Binishell—the dome shape is hemispheric and not elliptical, as in the more recent applications of the system.

Like the Australian multipurpose centre, this project is characterised by a cupola at the centre of the composition and additional volumes around it to conceal its connection to the ground. In this case, the contrast is between curves and lines; in other words, between the geometry of the sphere and the parallelepiped boxes that surround the Binishell. The only part of the structure visible from the outside is the entrance. This opening was cut using orthogonal planes, thus further accentuating the tension between curves and lines.

Although this preliminary project was not developed any further or even built, it is still a testimony of Bini's ambition and his high expectations for the future of his newly-filed patent. Given the geographical and temporal proximity with the Italian domes designed and built by Nervi, it should be no surprise that Bini imagined using his pneumatic construction technique for such prestigious applications as large public buildings (Fig. 4.23).

Fig. 4.23 Dante Bini, early Binishell concept, 1964–1965. Model (DB)

4.4.2 The Binishell Sinks into the Ground

Bini never abandoned the idea of concealing the connection to the ground and the openings of the Binishell and he revisited this idea in 1999–2000 to develop a design proposal for the new de Young Museum competition in San Francisco, California (Fig. 4.24). On that occasion, Bini conceived a building that can basically be intended as an underground architecture: the Binishell was almost entirely covered by the surrounding land, which had been artificially shaped to create access paths to the museum. Such paths were arranged radially around the centre of the dome and they defined the perspectives through which the visitors were meant to approach the building and discover its main attraction—the Binishell. Overall, this project provides further evidence of how important geometry has been for Bini and how he has used simple solid forms to enhance monumentality and symbolism in his designs. Although this competition project was not realised, it remains an essential statement of Bini's ideas about integrating concrete shells as parts of more complex spatial compositions, with applications ranging from housing to public buildings.

Renzo Piano is one of the many architects who uses spheres and other pure geometric solids in public projects as parts of a more complex organisation of interior and exterior spaces. For example, he designed, between 2000 and 2008, the California Academy of Sciences, which involves a complex intersection of geometric volumes. The main building is a straightforward parallelepiped, and the internal spaces were carved using spheres, which also control the direct sunlight, ventilation and the visual relationships between indoors and outdoors. As most architects do, Piano uses his projects to convey messages. Geometry is one of the ways of conveying

Fig. 4.24 Dante Bini, Binishell project for the new de Young Museum competition, San Francisco, California, 1999–2000. Model (DB)

such messages and is also a strategy to communicate the hierarchy of the building functions through symmetries and different scales. For example, in this case, the spheres represent knowledge and the fragility of planet earth.

The main difference between Bini's proposal for de Young Museum competition and Piano's design for the California Academy of Sciences lies in the recognisability and perception of geometry. Bini creates an entire building around the Binishell, without, however, touching the cupola with the surrounding volumes. He does not consider the shell as a mere structure but as an architectural element so pure and central to the composition that it cannot be carved or annexed with other forms. On the other hand, Piano favours blending pure geometries into a fluid wavy roof, thereby highlighting permeability and porosity between spaces.

Another important analogy between the California Academy of Sciences and Bini's proposal for the de Young Museum competition concerns the perception of levels, and horizontal surfaces in general. The green roof of Piano's project can easily be experienced as being the actual ground level. Visiting the academy also gives the impression of being underground because of the way natural light infiltrates thanks to the presence of a large number of skylights. The de Young Museum design proposal attempted to achieve the same result by creating narrow paths that led to the Binishell, the entrance of which remained concealed from sight. These are in fact visual strategies that are used to submerge a structure, without actually building underground.

4.4.3 Underground Binishells

Underground architecture is a fascinating topic because it involves inverting the way buildings are conceived and experienced. Normally, we first perceive a building through its façade, skin or envelope, which is the element that defines the volume of the artefact and allows how much physical space it occupies to be established. The tectonic qualities of the façade offer a sense of the building mass. The openings define its porosity and permeability. The overall composition of the building and its surroundings can either indicate or hide the entrance; hence, it can either guide its visitors through a journey on how to experience the building or leave them disoriented. However, there are no elevations in underground architecture, and only one part of the building cannot be hidden from external view: the entrance. The designing of such an opening is of course of utmost importance since it is the only element that relates this kind of structures to the outside world. It also defines the journey through the building because it determines how it is experienced.

When discussing underground architecture, it is not so natural to refer to conventional terms, such as wall or roof. However, it is possible to discuss such projects from the circulation point of view. For example, an underground building can be accessible from the roof by means of a horizontal skylight positioned at the ground level. This solution allows an underground structure to be experienced starting from the upper floors and moving down to the lower floors in a hierarchical order that is opposite

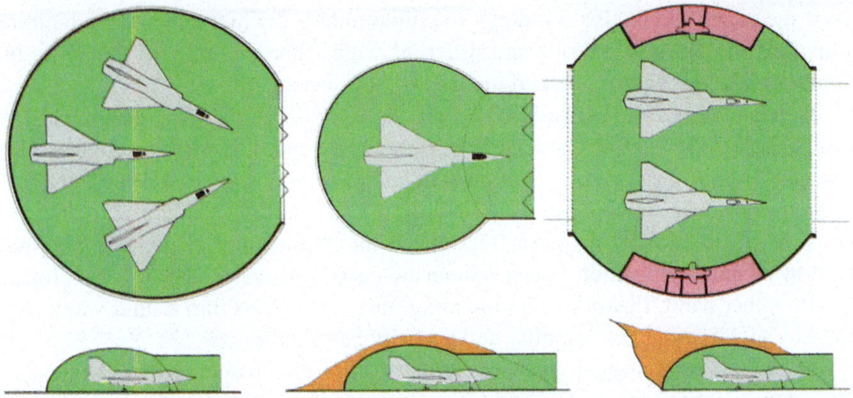

Fig. 4.25 *Binishells spa*, military shelters, late 1970s. Plans and sections (DB)

that of buildings constructed above the ground. However, a less extreme solution that can be adopted is to design a vertical opening, either at the ground level or via a ramp or a tunnel: this latter approach was used to conceive several Binishell hangars. It is reasonable to assume that certain functional requirements, such as the accessibility of vehicles, favoured this choice instead of an option whereby the section prevails over the plan, which would have required a sophisticated technology to enable the vertical circulation and transportation of people and goods. However, it is almost a pity that such a fascinating topic as underground architecture can mostly only be found in those Binishell applications that are merely technical, and unbuilt, such as the military shelters designed between 1984 and 1993 for the Guernsey island airport, located between the UK and France (Fig. 4.25).

4.4.4 The Binishell Flies

Over the years, Bini investigated several strategies to integrate a Binishell structure with its surrounding landscape. Moving in the opposite direction of submerging the dome into the ground or concealing it behind other building volumes, Bini studied the possibility of creating a buffer between the concrete shell and the earth.

In Styles' thesis, it is possible to find some of Bini's design explorations that belong to this category. For example, Styles included a sketch design for a Binishell sitting on a base, which raised the level of the structure above the ground plane. It also defined a threshold between public and private spaces, which was much more permeable and less demarcated than having a concrete dome directly touching the ground. In fact, it looks as if the Binishell was floating, and it seems much larger than the mere space created by the structure. This is due to the presence of a base and large openings that offered an opportunity to extend the interiors by connecting them to the exterior spaces.

A similar design approach was used for the 1977 Ashbury Public School multi-purpose hall in New South Wales. Even though this Binishell was not constructed according to the conceptual design drawings included in Styles' thesis, these preliminary sketches can be considered relevant because they highlight how Bini was attentive to the architectural integration of his structures in the surrounding landscape (Fig. 4.26). They also show that the initial design idea for the Ashbury Public School Binishell featured many more openings than were actually built. Such openings were also integrated with outdoor paths to guide the users towards the building from specific directions and angles. It is natural to relate such attention to the context with the lessons that Bini most likely learnt from his discussions with Michelangelo Antonioni during the design of his holiday home in Sardinia.

This sketch design also highlights that structural behaviour was not one of Bini's priorities; it was instead the integration of a shell structure within the landscape that mattered. There are many examples of large-span shell structures that use the connection to the ground as an opportunity to express, with elegance and lightness, how forces are transferred from the dome to the ground. For example, in the Palazzetto dello Sport by Nervi, there are clear visual pathways that can be followed. On the contrary, it is unclear, in Bini's design for the Ashbury Public School, where and how the dome touches the ground. Aesthetic choices seem to have been prioritised over structural concerns in the way of negotiating between architectural ideas and performance requirements that Robert Venturi would have defined as complex and contradictory.

Fig. 4.26 Dante Bini, Australian Binishell project, 1974. Preliminary sketch (Reproduced from Styles 1975: 127)

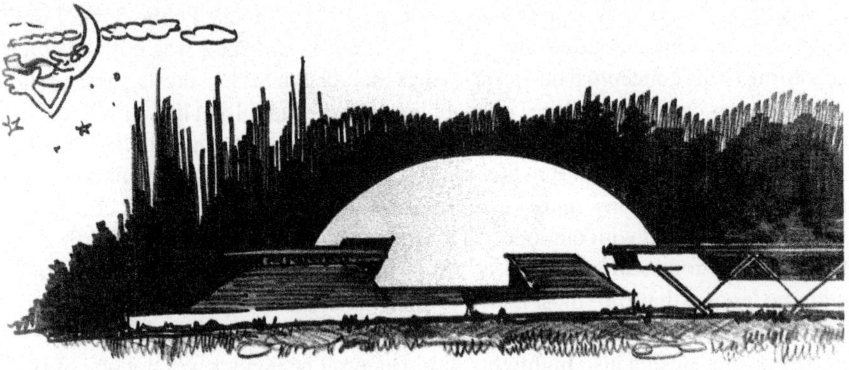

Fig. 4.27 Dante Bini, Australian Binishell project, 1974. Preliminary sketch (Reproduced from Styles 1975: 141)

From his very early designs, it is clear that Bini explored the tension between orthogonal and curved objects, almost as if to emphasise to what extent the curve belongs to the shell structure and the line to its surroundings. It is almost the opposite of the idea that modern architecture is mainly orthogonal and is located in a natural context of curved objects and complex forms. The Binishell is, to all intents and purposes, a curved but pure object that sits on another artificial and geometric element defined by lines. The aim of a Binishell is to be as permeable as possible and to take root—that is, be fixed to the ground like a tree. Ad-hoc designed outdoor paths can create this effect, radicating a concrete cupola to the site and further accentuating the difference between lines and curves (Figs. 4.27, 4.28, and 4.29).

In his thesis, Styles wrote about this design strategy by comparing the Binishell with a living cell and the outdoor paths with the arteries that feed it. It is an interesting organic analogy that could be used to analyse an isolated dome within its context: for instance, imagining the structure as a unicellular organism with a parasitic relationship with the land on which it sits. It might also work to describe clusters of Binishells, which is the topic of Chap. 5.

4.4.5 The Binishell on a Sloping Site

The outdoor paths of the Ashbury Public School somewhat recall the design of Villa Antonioni in Sardinia, which is accessed via a small footbridge that points towards the cupola and, as a consequence, to the sea in the background. However, in Villa Antonioni, outdoor paths play an even more important role, given that the house sits on an inclined hillside, and the entrance is located on the first of two floors. The paths are a way of solving the difference in height between the house and the access

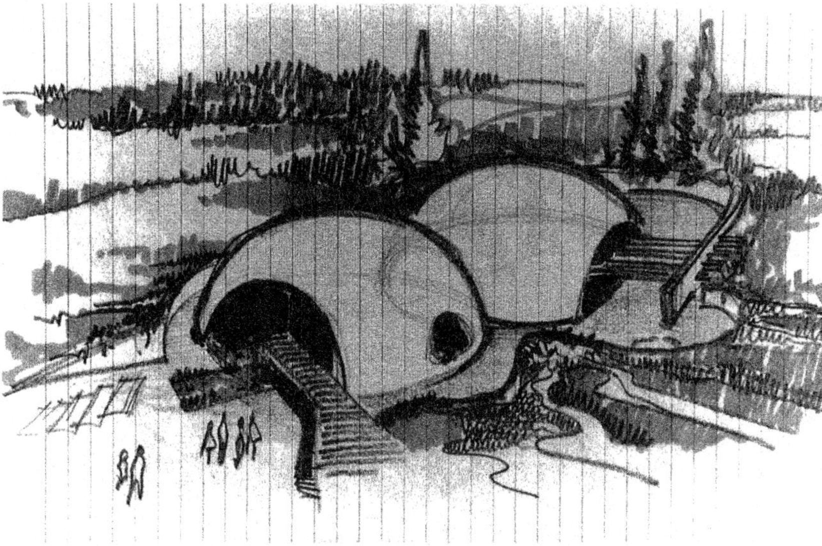

Fig. 4.28 Dante Bini, Australian Binishell project, 1974. Preliminary sketch (Reproduced from Styles 1975: 126)

Fig. 4.29 Dante Bini, Australian Binishell project, 1974. Preliminary sketch (Reproduced from Styles 1975: 127)

road. Once inside, the curvilinear openings are the strategy that Bini used to frame particular views of the sea.

It is possible to find several similarities between Villa Antonioni and the Bianchi House at Riva San Vitale, in Switzerland, designed by Mario Botta in 1971 (Fig. 4.30). Both projects are located on a steep slope, with the site's topography informing the house design. Both make use of pure geometries, in contraposition to the natural surfaces of the terrain. The internal spaces are created, and the views are framed via a subtraction of volumes, even though the Bianchi House employs orthogonal lines and parallelepiped solids, while Villa Antonioni is based on an elliptical cupola and curvilinear cuts on its surface. The Bianchi House is laid out on more levels than Villa Antonioni, but, in both cases, access occurs from the top floor via a narrow footbridge. The vertical circulation is arranged centrally to the plan, and its synergy with the openings generates a strong connection between indoor and outdoor spaces. The concept of the *promenade architecturale* was applied in both projects in a way that requires the inhabitants and visitors to take an outdoor path before entering the dwelling, therefore expanding the physical dimensions of the building through visual connections with the surrounding landscape.

Fig. 4.30 Top: Dante Bini, Villa Antonioni Binishell, Costa Paradiso di Gallura, 1969–1971. View of the entrance pathway (DRS); Bottom: Mario Botta, Bianchi House, Riva San Vitale, 1971. View from the entrance footbridge (AZ, MB)

Fig. 4.31 Dante Bini, Narrabeen North Public School, New South Wales, 1974. Sketch of the main Binishell opening, called the "pearl of knowledge" (DB)

4.5 The Shape and Size of Openings

Although sculptural and curvilinear openings began to appear in Binishells in the very early designs, the most popular and published project featuring such cuts is among the later examples that were built in Australia: the three domes completed in 1974 at the Narrabeen North Public School. A conceptual sketch that Bini made for this school project shows the shell surface being lifted—as if it were a curtain—opening its doors to young scholars and revealing the "pearl of knowledge".[1] Bini's drawing is a vignette that transforms the brutality of tearing down a part of the shell to make an opening into a poetic image: the pearl is a direct reference to the function of the building—as a library—and an indirect reference to the purity and simplicity of structural form (Fig. 4.31).

In his thesis, Styles compared the aesthetic qualities of curvilinear cuts to inserting orthogonal doors and windows in a Binishell. He argued that openings were rarely perceived as the architect drew them, that is, in elevation in concrete domes, adding that this is particularly true when such openings were not designed using orthogonal lines. At first, this might sound like an obvious statement since an orthogonal projection—in this case, an elevation—is quite different from a perspective view, which is what the human eye sees. But Styles' comment becomes clearer when considering

[1] The "pearl of knowledge" is an expression that Bini has recurrently used when describing this project to the book authors.

that our point of view is not static and keeps changing as we move. Although we can somehow relate plans and elevations to what we see when walking around buildings that are mainly orthogonal, things are more complicated with curved shapes. In a Binishell without openings, concepts such as façade, front, back, sides, corners, walls and roof, have little or no meaning. A single elevation is sufficient to describe its shape, and a walk around the structure can even disorient us, given the absence of formal variations.

Puncturing such a pure form with standard doors and windows would allow the position of the entrance, the sides and the back of the building to easily be determined, but—architecturally—it would also be a way of ignoring the presence of a concrete cupola whose aim is not that of simply being experienced as a structure with straight walls and four elevations (Fig. 1.25). A number of Binishell projects built in Mexico by the *Binica sa* feature large square and rectangular openings, and clearly illustrate the risks of adding orthogonal lines and planes to a shell form. Instead, designing such a structure with curvilinear cuts creates a condition in which the perception of the surface of the cupola changes as the observer's point of view varies (Figs. 4.32 and 4.33).

The entrance opening of one of the Narrabeen Binishells, with its curved profile and the pearl of knowledge supporting the structure above the cut, is probably the most iconic image of Bini's projects (Fig. 4.34). It has been documented with myriads of photos, taken from many different angles, which demonstrate how critical the relationship between curvilinear cuts and the human experience of a shell surface can

Fig. 4.32 Felipe Humberto Concha Hernández *(Binica sa)*, Telephone Exchange Binishell, Mexico City, 1970s. External view of the entrance (DB)

Fig. 4.33 Felipe Humberto Concha Hernández *(Binica sa)*, Telephone Exchange Binishell, 1970s. Drawing (DB)

be. For example, a picture taken from the southeast highlights the pearl of knowledge under the Binishell, while a photo from the southwest primarily shows the wide curved cut, leaving the sphere underneath mostly in shadow (Fig. 4.35).

The shape of the cuts is only one of the variables that can be investigated when studying openings in a Binishell structure. Scale and size are other important factors,

Fig. 4.34 Dante Bini, Narrabeen North Public School, New South Wales, 1974. The entrance of the library (DB)

Fig. 4.35 Dante Bini, Narrabeen North Public School, New South Wales, 1974. Elevations (DB)

as well as the relationship between the position of the opening and the arrangement of the internal functions.

For example, in many of the design sketches prepared for the Australian projects, and published by Styles in his thesis, it is possible to notice that the curvilinear openings work in synergy with the layout of the external paths to give the impression of the building expanding into the surrounding landscape. A similar approach can also be found in the project for Villa Antonioni, where the openings define the visual and physical communication pathways between indoors and outdoors, radicating the building to the site.

An experimental 20-m diameter Binishell designed by Furio Nordio—advertised via a *Binishells spa* catalogue in May 1967, but which was never built—was conceived following a diametrically opposite strategy (Fig. 4.36). The shell features one main cut, which was made particularly wide in order to be able to trim the apex of the dome. On the one hand, this large hole allowed the architect to create two windows facing onto an internal courtyard; on the other hand, the shell seems to have lost its primary structural function, thus becoming a mere visual shelter that guarantees a certain degree of privacy to the inhabitants. The absence of a context in which to place this conceptual design further emphasises the contrast with Villa Antonioni, where the cupola is fully integrated with its surroundings and acts as a tiny element of a much larger system.

In his thesis, Styles included several images of Binishell projects that have never been built. In many instances, neither the name of the architect is specified nor the origin of the drawing. In some cases, the information provided in the caption does not even correspond with the description of the same project found elsewhere, such as in the many *Binishells spa* promotional catalogues. But despite this difference, some of these images show approaches to the design of openings that are somewhat unique. For example, a sketch design—which illustrates a self-contained Binishell

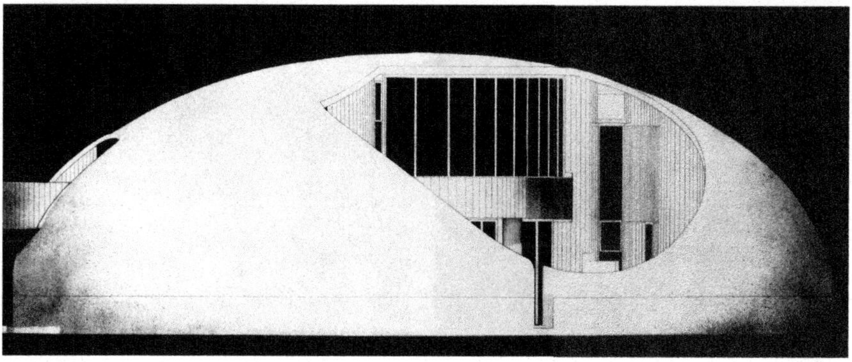

Fig. 4.36 Furio Nordio, experimental Binishell house project, 1967. Elevation (DB)

restaurant with large cuts and an internal courtyard—can only be associated with Nordio's experimental house project discussed above.

When the size of the cuts is particularly significant, it is worth pondering on whether it is the Binishell system that has shaped a new concept of dwelling or, vice versa, it is the traditional rules of housing that have eroded the structure up to the point that is has become an ornamental element of the composition.

4.6 Generating the Openings During Inflation

The need to cut openings on a newly erected Binishell is the construction and structural problem that has inspired the topic and narrative of this chapter. However, not all systems that Bini developed to build concrete shells using air required the openings to be obtained by means of subtraction from an already inflated structure.

A saddle-like concrete shell prototype—which Bini had already made in 1967 as part of his early experiments, undoubtedly inspired by Heinz Isler's and Félix Candela's designs—is one of these cases (Bini 2014). The membrane was shaped in such a way as to concentrate the fresh concrete within a specific boundary on its surface, thus creating openings in the remaining areas during the inflation process. The photographs taken during the construction of this shell prototype illustrate how Bini exploited the curvature inversion principle to design a membrane that, at the same time, defines structural form and the shape and size of the openings (Fig. 4.37).

Although this experiment did not lead to the development of a specific patent for inflatable shell structures with anticlastic curvature, the idea behind the prototype was nevertheless integrated into another system: the 1979 Minishell. As discussed in Chap. 3, the Minishell can be considered as an evolution of the Binishell. It is a system that is based on a square footprint, and it was conceived to provide an alternative to the circular plan and structural form of the Binishell for residential applications. The shape of a Minishell is closer to the volume of a box, thus making it easier to

Fig. 4.37 Dante Bini, saddle-like concrete shell prototype built at the "mushroom field", San Cesario sul Panaro, 1967. Inflation process (DB)

use for small housing projects. The membrane used to erect a Minishell also creates four openings at the corners of the structure. Unlike the original Binishell system, it does not require cutting the openings after inflation.

In 1980, five Minishells were constructed by *Jennings Industries Ltd* as holiday apartments at Trinity Beach, Cairns, in Queensland. By comparing these structures with their Italian counterparts, which were based on the older construction system, such as the domes built at the Torre Cintola holiday resort, it emerges how the cut of the openings in Bini's residential projects always tended to square the curves of the shell to satisfy certain functional requirements. In the Torre Cintola Binishells, the cuts were made with vertical planes to create orthogonal wall portions and to be able to use standard door and window components (Fig. 4.38). These lines also helped reshape the circular perimeter of the structure in order to rationalise the arrangement of the internal functions. In the Minishells at Trinity Beach, the openings positioned at the four corners of the plan allowed those parts of the structure where height and curvature would have been an obstacle to the design of the internal layout to be eliminated (Figs. 4.39 and 4.40).

As mentioned in Chap. 6, Nicolò Bini has been actively working on developing new ways of automating the construction process. Many of his recent experiments and systems can be considered declinations of the ideas already implemented in the 1979 Minishell patent, i.e. creating the openings during the inflation process of the structure and adapting the shell plan to a functional or desired arrangement of

Fig. 4.38 *Binishells spa*, Torre Cintola holiday resort, Monopoli, 1970s. External view of a Binishell holiday house (DB)

Fig. 4.39 Dante Bini, Minishell holiday houses, Trinity Beach, Cairns, Queensland, 1979–1980. Construction sequence (DB)

the internal functions. Particularly on the issue of the openings, it makes sense that Nicolò Bini's systems be more inspired by the Minishell than the Binishell, given the growing awareness on environmental sustainability in the construction industry. Nowadays, it would not be conceivable to cut the openings from the finished concrete shell, thus using extra material and generating more waste.

Nicolò Bini favours nature as a design precedent rather than classical architecture. For example, in the villa he designed in Malibu between 2012 and 2017, which is

Fig. 4.40 Dante Bini, Minishell holiday houses, Trinity Beach, Cairns, Queensland, 1979–1980. Internal view of the raw shell structure (DB)

discussed in detail in Chap. 6, he positions the house openings at the base of the shell so that the natural light can wash the internal surfaces of the structure, emphasising its double curvature and the relationship with the external landscape. The finished concrete structure, before placing the internal partitions and the window systems, immediately recalls the image of a cave or similar natural form obtained over the centuries by erosion. On the contrary, Dante Bini privileged classical antecedents and symbolic images to find inspiration for his designs. For example, he employed skylights that evoked the oculus of the Pantheon several times to emphasise monumentality. He also gave shape to concepts, such as the "pearl of knowledge", which hides behind a sinuous cut on the Binishell surface of the Narrabeen school.

Chapter 5
Lunar Bases on Earth: Intersections and Repetitions

5.1 Issues and Opportunities with Modular Structures

The Binishell system was invented with the aim of simplifying and speeding up the construction of shell structures by using a pneumatic membrane to shape fresh concrete and bend reinforcing rods. It is an idea that blends three main concepts: standardisation, modularity and prefabrication (Fig. 5.1).

As demonstrated by the early prototypes and promotional catalogues produced by the *Binishells spa* company, the system proved successful for several residential, commercial, industrial and military applications. However, Bini soon realised that building single-dome structures would make his system unsuitable for larger-scale projects. For example, a single dome would not be large enough to cover an indoor sports hall that included a seating area. A single Binishell would also be too small for large shopping centres and schools. Bini responded to this challenge in two main ways. First, he used intersections to design large structures composed of smaller domes. This is not an uncommon design response and has been explored by many architects and engineers known for their contribution to the field of shell and spatial structures. For example, Félix Candela mainly used intersections of geometric forms in his projects, particularly hyperbolic paraboloids (Garlock and Billington 2008). Eduardo Torroja and Anton Tedesko also privileged geometry as a strategy for structural design, realising several projects based on the repetition and intersection of synclastic surfaces, such as cylinders, but also anticlastic surfaces, as Candela did with hypars (Torroja 1958; Engel 1997).

Bini used repetition as a second strategy. In this case, the difference between the concepts of intersection and repetition is subtle, and it is possible to argue that there are several overlaps between the two. Conceptually speaking, intersection concerns those Binishell structures that are trimmed to be physically connected, while repetition does not necessarily involve physical contact between building units. This is relevant from an architectural point of view because intersections primarily affect the design of the roof structure and—in some cases—of the programmes underneath

A. Pugnale and A. Bologna, *Architecture Beyond the Cupola*,
Mathematics and the Built Environment 7,
https://doi.org/10.1007/978-3-031-26735-2_5

Fig. 5.1 *Binishells spa*, Rice Export Corporation Pakistan Binishells, Pipri, Karachi City, late 1970s (DB)

this larger-span shelter. Repetitions do not simply inform spatial planning; they can also involve the landscape design of the outdoor spaces in between shells or the addition of other building elements that connect the structures, such as passageways and arcades (Fig. 5.2).

After a brief introduction to standardisation, modularity and prefabrication—and how these three concepts are related to the design of Binishells—this chapter illustrates how Bini used intersections and repetitions to expand the scope of application of his system, which ranges from holiday resorts to industrial applications. Several projects are discussed, including two of Bini's iconic Australian designs: the Narrabeen North Public School in New South Wales and the Space City Shopping Centre in Queensland.

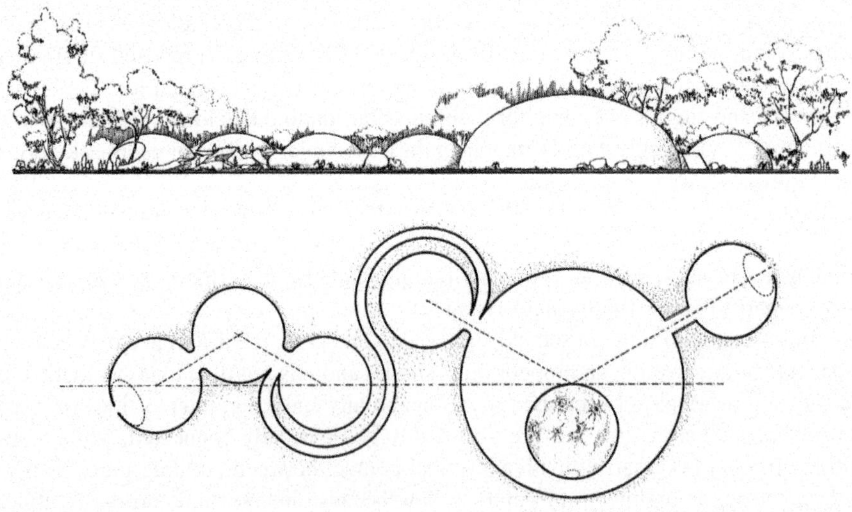

Fig. 5.2 Dante Bini, Binishell project for an entertainment park in the Union of Soviet Socialist Republics (USSR), 1987. Plan and Section (DB)

5.1.1 *Identity and Diversity*

The concept of modularity is not new to the field of structural design. Modular structures have existed since ancient times. During the Roman empire, many architectural masterpieces were built by repeating the same elements—primarily an arch—with the aqueduct of Segovia and the Colosseum in Rome being prominent examples. In the past century, modularity was extensively explored, particularly in the 1960s and 1970s, in relation to prefabrication and residential architecture. Moshe Safdie's Habitat 67 is one of the iconic projects of that period that made of identical prefabricated concrete forms an architectural feature (Legault 2021). The project can be described as an arrangement of parallelepipeds stacked in different ways to create apartment units that are unique, despite being derived from the same modular element. The final composition is so intricate that it is impossible to decipher its generative rules without referring to the drawings. It is an intended effect that generates a tension between apparent randomness and the intrinsic regularity and rationality of a modular prefabricated building. It is also a common design approach that values diversity and uniqueness over the rules of serial production.

The reason behind emphasising individuality and the sense of identity is related to the psychology of the habitation and appropriation of a house. One of the main issues of modular houses is the obvious reference to industrial elements, such as the container or the trailer, which deviate a great deal from the common idea of a dwelling. No one wants to live in a trailer: therefore, the designer's challenge refers to psychology rather than to the performative aspects of the modular structure design. According to Paul Rudolph, designer of the 1968–1970 Oriental Masonic Gardens in New Haven, Connecticut, the challenge is to provide an alternative to terraced houses or single-family houses, and to give each unit its own plot of land and individual courtyard (Fig. 5.3). It is a matter of arranging the modular elements so that, even though they formally recall the serial unit that originated them, they still encapsulate features of the dwelling (Rudolph 1970: 218–219).

The aspect of identity generation through diversity—in other words, arranging and assembling standard modular structures in ways that define unique spaces and forms—is critical. The shape of a Binishell is emblematic, and because it is curved, it is hard to recombine it geometrically with other similar or identical forms. This chapter explores how Bini responded to this challenge in a number of residential and non-residential projects.

5.1.2 *Flexibility and Adaptation*

Unlike Habitat 67, the 1970–1972 project by Kishō Kurokawa for the Nagakin Capsule Tower in Tokyo is a representation of its industrial and technological nature

Fig. 5.3 Paul Rudolph, Oriental Masonic Gardens, New Haven, Connecticut, 1968–1970. Bird's eye view of site (LC, PRC)

(Kurokawa 1977: 105–111). The building is composed of capsules that are independently attached to a steel and concrete frame. In this case, the architectural themes explored prioritise—for example—flexibility, adaptation and replacement.

Between 1970 and 1971, Renzo Piano and Richard Rogers explored similar concepts of flexibility, adaptation and expansion over time for the Centre Pompidou in Paris, an iconic project that made of the structure and the services the aesthetics of the façade so as to liberate the interiors from physical constraints, thus ideally allowing the many uses of the museum spaces to be changed over time (Dal Co 2016). The two architects worked extensively on the concept of flexibility during the early years of their careers, realising many original projects. For example, the 1971 B&B Italia Offices in Novedrate encapsulates a smaller-scale example of the concepts explored for the design of the Centre Pompidou. They designed and built single-family houses in Cusago, near Milan, between 1970 and 1974, following an industrial logic, with a truss roof structure that liberates the interiors from intermediate supports by transferring all the loads onto the two facing sidewalls (Buchanan 1993: 50–51). This strategy also allows the house to be expanded in size over the years since the other two walls are glazed, and the structural bay can easily be repeated.

Flexibility is another critical aspect of this chapter, since it seems to be the underlying concept used to develop—and to promote—many Binishell school projects and holiday resorts as being innovative. The idea behind flexibility in architecture is that a large building complex can be created by repeating the same structure a number of times until the expected size is achieved. In the design of Binishells, the individual domes were generally joined by adding conventional building elements, such as passageways and arcades. This concept of flexibility also includes the possibility of future expansions. In most Binishell designs, the finished project only represents one of the possible outcomes of the same idea, where it is the number of repetitions that

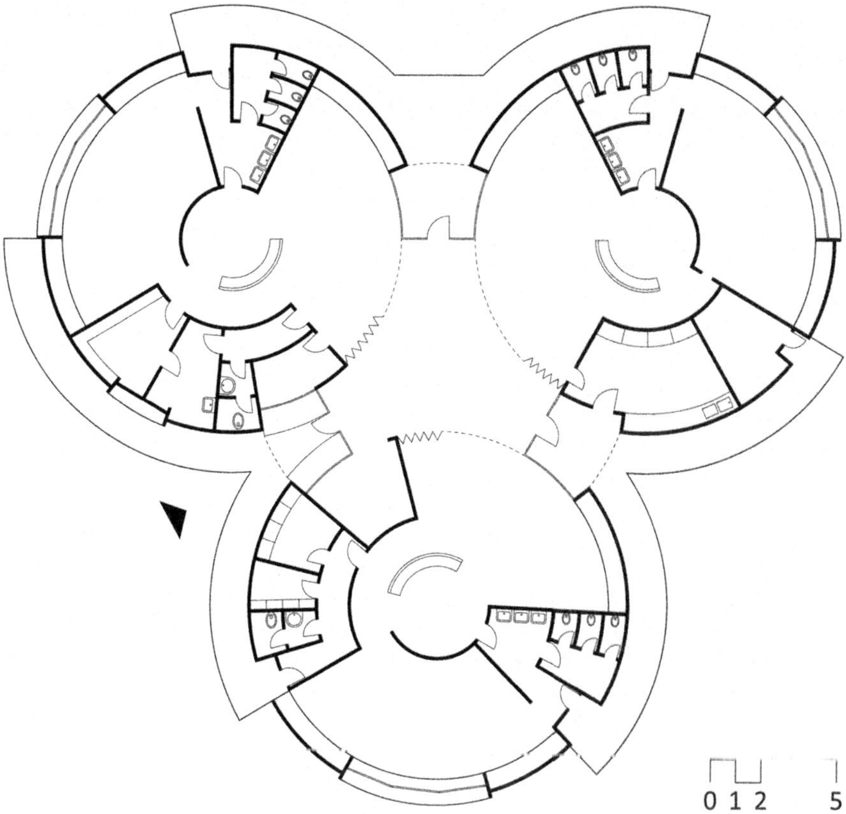

Fig. 5.4 *Binishells spa*, three-section school project, 1970. Plan (AP)

identifies the final scale. For example, a three-section school made of three domes can be expanded in the future to accommodate more students by adding further domes to the system. Another example is provided by a holiday village project that could potentially expand over time. It is a design logic of repeating and joining shell structures that adapts to the needs of different clients and which can be readapted in the future, if the conditions change (Figs. 5.4 and 5.5).

5.1.3 Structural Rationality and Simplicity

Geometry rationalisation has always been an important design aspect in shell and spatial structures since construction efficiency, cost and the tectonic qualities of the building depend on it. It was even more so before the development of digital tools for the design, optimisation and fabrication of complex forms. Pier Luigi Nervi is probably the most notable example of a designer who made an art of prefabrication

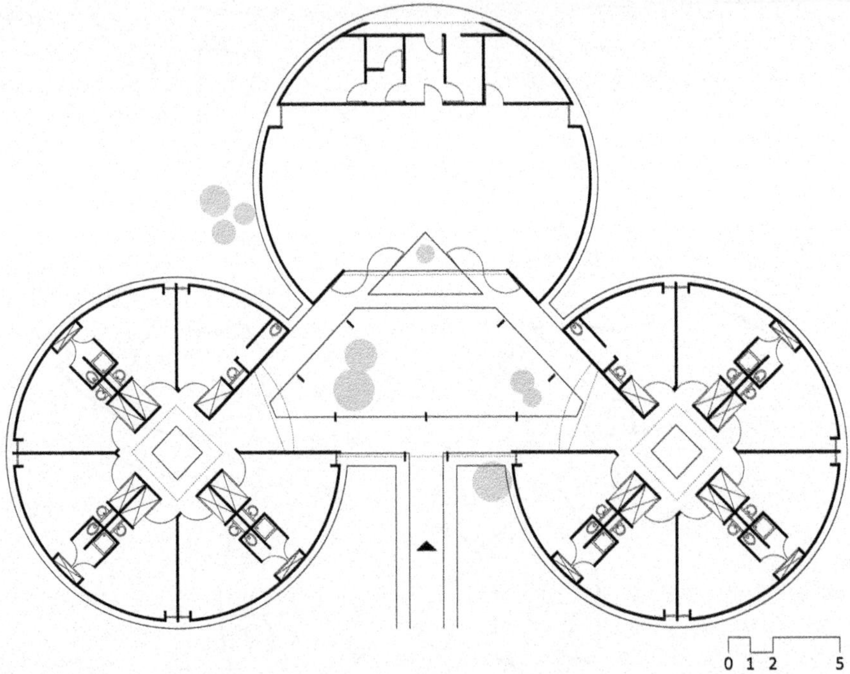

Fig. 5.5 *Binishells spa*, holiday resort project, 1970s. Plan of a three-dome module (AP)

techniques to erect buildings, from relatively simple geometric shapes and the repetition of patterns, that could be built on-site efficiently. Nervi's idea was to emphasise geometric clarity, structural lightness and how the forces run through a structure (Gargiani and Bologna 2016).

Structural rationality and simplicity are relevant in this chapter because Bini primarily used prefabrication and geometry to solve architectural and construction problems rather than structural issues. There is only one documented example that partially goes against this logic—a prototype for a shell roof component for industrial applications—which is introduced in Chap. 1 and is further discussed below. It is worth noting that this is the only case in which Bini focused on designing a serial component rather than a finished building made of repeatable elements.

5.2 Repeating Binishells

In his designs, Bini responded to the many challenges described above in a variety of ways, depending on the type of project. The concept of uniqueness and identity were generally prioritised for residential housing applications. He focused more on technology for public and commercial buildings, and in particular on the flexibility of the system that was to be reused for different purposes. Repeating the same element is one of the characteristics that makes the Binishell system fast and affordable. Therefore, Bini had to deal with repetitions and intersections as the two main design strategies.

In some instances, the Binishell system mutated into something else to respond to a specific design requirement. For example, the 1979 Minishell is a pneumatically erected concrete structure with a square footprint and openings at the four corners. It can be considered an evolution of the original Bini's construction method because it eliminates the issue of the circular plan used for residential applications (Figs. 3.29, 3.30, 4.39, and 4.40). Even though it would be quite natural to trace the history of Bini's systems in chronological order, that is not the aim of this chapter, which is focused on design strategies rather than on building systems and techniques. Therefore, shape variations are discussed as possible ways of composing and articulating domes in a project rather than being seen as technological evolutions of an invention.

5.2.1 *From the Casa Cupoletta to Holiday Resorts*

The repetition of modular elements was essential in such residential projects as Habitat 67 or the Oriental Masonic Gardens to obtain certain apartment sizes. Therefore, generating a sense of identity and individuality was related to the need to assemble smaller building parts to form a whole that could satisfy specific programmatic requirements. The process of composing and articulating identical objects provided a direct opportunity to solve one of the architectural issues of designing houses with modular and prefabricated elements.

Such a process works somewhat differently for residential Binishells since a 12-m diameter dome can provide shelter for a single-family house of roughly 110 square metres. Here, the concept of repetition is related more to reducing construction costs. Reusing the same pneumatic membrane and replicating the same shapes a number of times makes the Binishell system affordable and efficient. However, architecturally speaking, building more than one concrete cupola is not a requirement to design a decent-size house. This argument is no longer valid when discussing industrial, commercial or other public applications. Still, it is possible to claim that—for residential buildings—the challenge of creating a sense of identity and individuality is not related to the serial production provided by the Binishell system. It is also true that the idea of living in a cupola is certainly rare among contemporary Western people. It is fair to conclude that transforming a structural dome into a house is the main challenge when designing residential Binishells. This is a typological problem that preceded the use of modularity and prefabrication. Therefore, it even anticipated the concept of repeating the same design multiple times.

Luigi C. Olivieri's Casa Cupoletta, previously mentioned in Chaps. 2 and 3, encapsulates all the basic stylistic features of a single-family house: it is immersed in green; it has a barbecue area and a pergola; a tall, prefabricated chimney; doors and windows are cut from vertical brick walls. Even though the Binishell was described as an igloo or a tennis air dome by the architect himself—and this was rather unusual for a house design—the project was not advertised by leveraging on its unique shape. It was all about the technical innovations that were used to reduce the costs, while preserving the characteristics of the prototype of a single-family house. As illustrated in Chap. 2, the main photograph in the *Grazia* magazine article shows Casa Cupoletta immersed in an isolated green lot (Fig. 2.16). The house is behind a tree and is painted in a light-yellow colour, most likely to reduce the perception of living in a concrete structure. The interiors are described as spacious, but no mention is made of the curved ceiling, except that this shape leaves room for services and the heating system.

The concepts of modularity and prefabrication can only be applied successfully on the housing market if the result matches the expectations of the potential owners. The Binishell shape was not the key selling point and neither was its technological advancements. It is clear that Bini was well aware of to what extent creating a sense of individuality and identity was essential for the commercial success of his invention.

At present, finding out how many of the Casa Cupoletta projects were constructed around Italy seems an impossible task. However, fortunately, we know that a similar design was used to build various small Binishells for the Torre Cintola holiday resort in Monopoli, in the 1970s (Fig. 5.6). By studying this project, it clearly emerges that a number of different architects who designed Binishell houses considered modularity as a design challenge of this construction system: therefore, the problem was primarily related to the formal features of a concrete dome.

The Torre Cintola holiday resort comprises forty identical single-family cupola-houses located by the seaside, arranged in circles, in three groups of eight, twelve and fourteen. Even though the Binishells are physically close, they neither touch nor intersect. They are stand-alone buildings but function like terraced houses. As in the case of Casa Cupoletta, the selling points are the proximity to nature, functionality, independence, but also luxury. Repetition is a mere technicality in this design, and it allowed the builder to reuse the same pneumatic membrane many times to optimise construction. Two larger Binishells of different sizes provide shelter for the reception and facilities of the resort. This is another case in which the shells do not intersect, even though they are connected, and their functionality is expanded through the addition of other more conventional building elements (Fig. 5.7).

In this project, the intersection was not explored as a design strategy, and repetition was used in a very pragmatic manner. Still, Torre Cintola can be a useful case study to highlight some of the limits of the original Binishell system—for example, the circular plan, which was later transformed into an opportunity in future projects by changing the shape of the shell or integrating the landscape as part of the design (Fig. 5.8).

5.2.2 A Landscape of Binishells

Spending a few days at Torre Cintola is surely an unforgettable experience: it is geographically located in one of the most beautiful spots in southern Italy, it is close

Fig. 5.6 *Binishells spa*, Torre Cintola holiday resort, Monopoli, 1970s. Page from a promotional catalogue (DB)

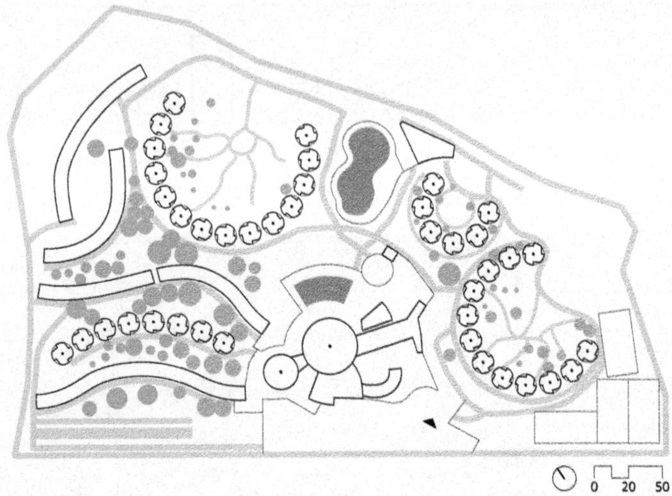

Fig. 5.7 *Binishells spa*, Torre Cintola holiday resort, Monopoli, 1970s. Site plan (AP)

Fig. 5.8 Alexandre Jeleff, holiday house project, Guadeloupe, 1969. Site plan (AP)

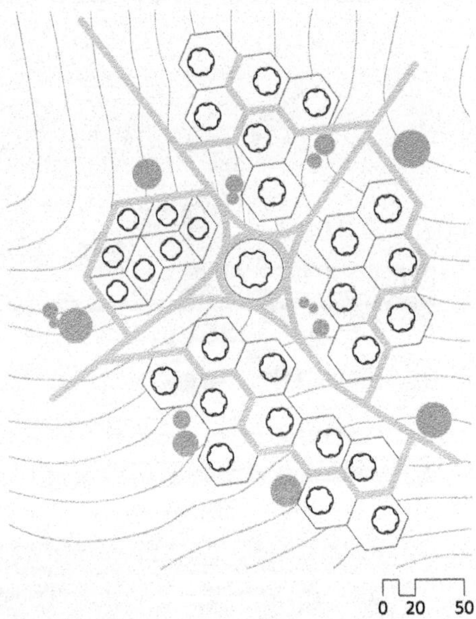

to the beach and it is immersed in nature. Even though the building surroundings are artificially recreated to look natural, visitors could think that the forty-two Binishells look like a lunar base in an untouched piece of land. The contrast between natural and artificial is marked and can be seen an integral part of the beauty of the place.

From the architectural point of view, the holiday village that Bini designed at the Cappuccini island in 1970, in Sardinia, shows a much more attentive approach to the context than Torre Cintola for a few reasons. After completing Villa Antonioni, Bini was able to secure new clients on the island, and this project arose as a result of the new contacts Bini made through Michelangelo Antonioni. This is a smaller village comprising eight residential Binishells. Unlike Torre Cintola, the contrast between natural and artificial is not as striking, primarily because the eight domes are cladded in stone, even though they are visually and physically independent from one another. It is also worth noting that the village can be seen from the sea when travelling to and from the island. This is an unusual point of view but an important one for an architect who is about to design a sort of lunar base on an untouched island. Unfortunately, the most relevant aspect of this project—the feature that makes it more advanced and architecturally appealing than Torre Cintola—was not built. Initially, Bini envisaged a tensile structure to visually link the eight Binishells and cover the outdoor spaces in between the houses. This structure was supposed to have been shaped by two large masts located in the centre of the village and by a number of support points at the ground level around the perimeter. According to Bini, the tensile structure was not built for a combination of reasons, all related to costs. Frei Otto was contacted to engineer the design, but his fee was too much for the budget. The same applies to the construction costs. However, a model and sketches of this idea have survived, and they clearly demonstrate that Binishells and tensile structures can coexist in a unique composition of curves, in both plan and elevation, blending the natural with the artificial (Figs. 5.9 and 5.10).

Fig. 5.9 Dante Bini, holiday village, Cappuccini island, 1970. Sketch (DB)

Fig. 5.10 Dante Bini, holiday village, Cappuccini island, 1970. Model (DB)

This project, which was carried out after Villa Antonioni, provides further evidence of how relevant the landscape was for Bini in his designs. Even though the shape of his system leads to a visual friction between the natural environment and the mushroom shape of the concrete cupola, Bini explored several different ways of blending his structures into the surroundings. Otto used tensile structures in a similar fashion in a few projects, including the German Pavilion at the 1967 International and Universal Exposition in Montreal and the 1972 Olympic Park in Munich (Otto 2005). This particular solution is relevant here because it illustrates how eight identical Binishells can become a whole organic system thanks to the addition of a tensile structure and a thoughtful design of the outdoor spaces in between the shells.

5.3 The Building and the Mushroom Field

The early Binishell prototypes were built in the "mushroom field" in San Cesario sul Panaro, near Castelfranco Emilia. It is intuitive to imagine how the name of the field originated: this pneumatic construction system allowed Bini to literally make Binishells spring up like mushrooms. As described in Chap. 1, shaping concrete dome structures through inflation is and was faster than any other shell construction technique. In fact, a Binishell can be erected in just a few hours in one single day. It would certainly have been incredible to come back home from work in the evening and see a new structure along the road that had not been there in the morning.

Reusing the inflatable membrane more than once was also essential to ensure the entire operation was financially viable. There are costs associated with making a single dome that do not necessarily make this technique cheaper than other more conventional ones. However, building several domes—therefore recycling the same membrane and reusing the same pumping station and associated tools—is faster and more efficient. This is a way for a builder to save money, and it explains why many of the projects presented above consist of more than one Binishell; even projects for single-family houses were designed in the kit-home fashion, with the idea of selling a readaptation of the same project to different clients.

In light of this, each Binishell project can be seen as the sprouting of a mushroom field. These are projects that go beyond the realisation of a single building because they embed the concepts of serial production, modularity and prefabrication. In Chaps. 3 and 4, Binishells were primarily discussed at the building scale, by analysing plans, sections and openings. Here, the focus is on how a single concrete dome can "proliferate" to form a larger and more complex commercial, industrial or public building.

The reference to mushrooms recalls the concept of growth, which is useful to describe, for example, how many Binishell school projects were generated with the potential of expanding them over time. The term "building" represents a conventional and ordinary building—such as a box—in contraposition to Binishells, which are the "mushrooms"—the foreign bodies. It is essential to emphasise such a difference because one of the strategies used to compose these concrete structures through operations of repetition and intersection involves the addition of more conventional architectural volumes, primarily boxes and geometries, such as cylinders and pyramids.

5.3.1 Binishells, Galleries and Arcades

When Binishells do not intersect—and are not designed to work as separate entities—they are generally linked through another ordinary volume. The aquatic centre in Arezzo, which is still in business under the name "River Piper Aquapark", is an example of this kind of approach that can be found in many *Binishells spa* promotional catalogues and advertisement material (Figs. 5.11 and 5.12).

This is a simple project that consists of two dome structures of different diameters, that is, 15 and 20 m. The two concrete shells are trimmed by means of vertical planes, which are almost perpendicular to each other and generate large openings that face each other and a swimming pool. The dialogue between the two structures is further emphasised by the presence of a one-story glazed volume, which looks like a narrow gallery that penetrates and connects the Binishells. In plan, this element is much larger than a simple corridor. However, it appears as a secondary architectural body in elevation, hierarchically speaking, since it is fully transparent and is not as tall as the shell arch openings.

The original aspect of this project lies in the presence of these almost perpendicular vertical openings, which—together with the gallery-like structure—define a façade such as the type that can be found in any other ordinary building. In this case, the permeability between the indoor and outdoor spaces is a huge difference, as the openings and the glazed gallery volume avoid that sense of barrier given by the concrete shell structure. The two large arches give a sense of openness and

Fig. 5.11 *Binishells spa*, River Piper Aquapark, Arezzo, 1968–1969. External view (DB)

Fig. 5.12 *Binishells spa*, River Piper Aquapark, Arezzo, 1968–1969. Site plan (AP)

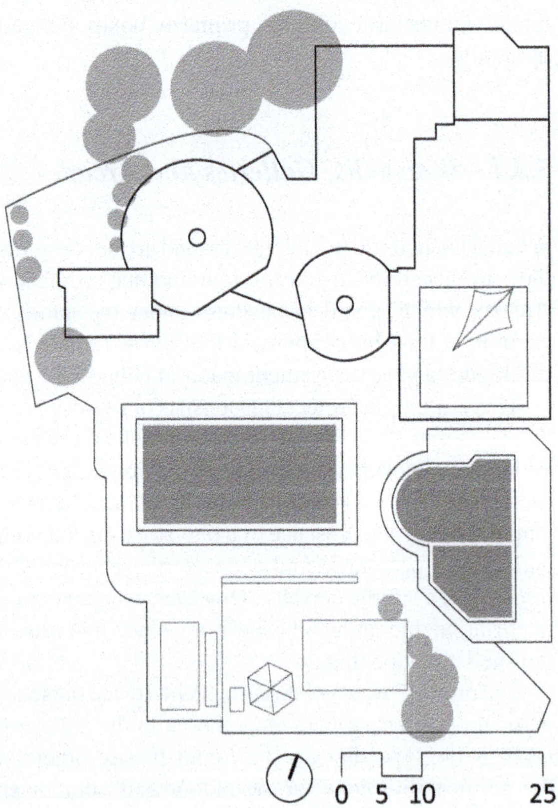

accessibility, and they also allow the shell thickness to be seen, therefore increasing the perceived lightness of the two cupolas.

A school designed by the architect Riccardo Merlo in Bologna is another example of Binishells linked via additional orthogonal volumes. This is a more complex project that comprises eleven shells—of which one large 30-m structure is used as the sports centre and administration offices, and the other ten 20-m domes are used as classrooms and a dining hall. The 30-m dome is the fulcrum of the composition, and the classrooms linked by galleries develop linearly on two opposite sides of this central space (Fig. 5.13). This large structure is the only one that was built as planned, while more conventional boxes replaced the smaller foreseen Binishells in the final project. However, by focusing on what was conceived on paper, it emerges that the main idea was repetition, with the added concepts of growth and flexibility of the school plan layout. The galleries used to connect the domes seem to suggest the school could have expanded over the years to accommodate more and more students and activities just by extending the galleries and adding further Binishells along them.

The same concepts were used in the project for a motel and service station designed by the architect Anna d'Alessandris Pazzi and the engineer Angelo Berardi. According to a *Binishells spa* promotional catalogue from that period, dated 1969, where this design is illustrated (Figs. 5.14 and 5.15), the project included two large 30-m and 40-m diameter domes, which were to function as the motel reception and restaurant, respectively, plus 15 smaller 10-m and 15-m diameter domes, where the motel rooms were to be located. The layout is conceptually similar to the day nursery in Rezzato, in which a large Binishell functions as the fulcrum of the project while the smaller concrete shells develop around it. This is another case in which the layout

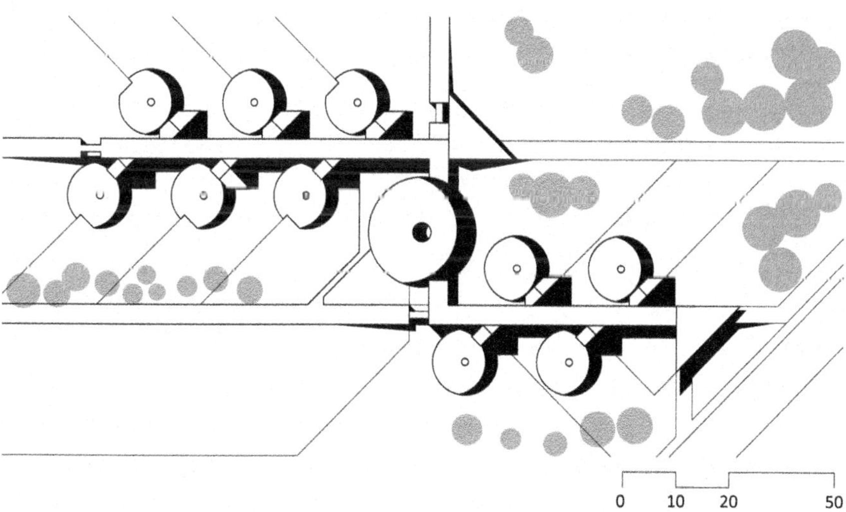

Fig. 5.13 Riccardo Merlo, Binishell school project, Bologna, 1969. Site plan (AP)

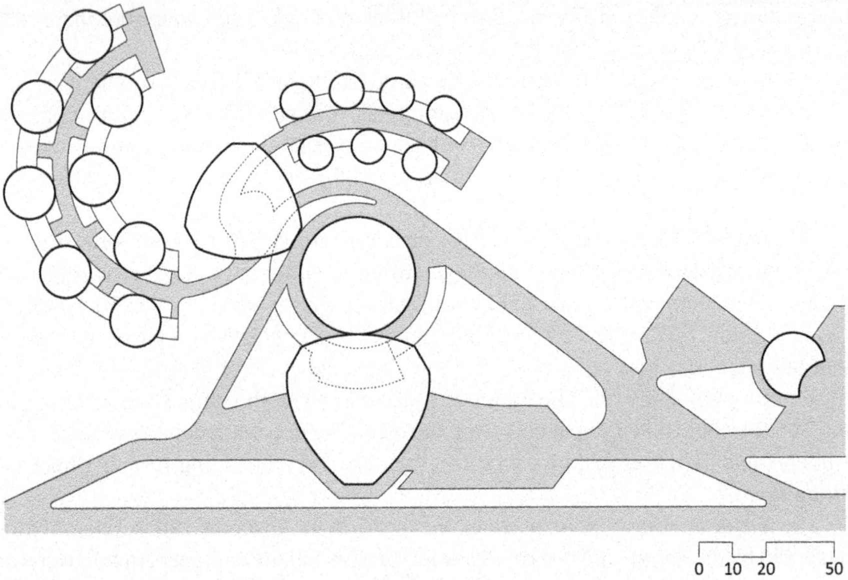

Fig. 5.14 Anna d'Alessandris Pazzi and Angelo Berardi, motel and service station project, November 1970. Site plan (AP)

of Binishells connected by galleries suggests the complex could expand over time. Construction-wise, it is worth noting that the small domes are all identical, thereby allowing the building company to save on costs through the repetition of modular elements. Instead, the two large concrete cupolas are presented in the promotional catalogue as the two distinct design features of the project.

Many design precedents come to mind when reviewing these Binishell design proposals, and it is probably not by chance that such projects were also developed in the late 1960s. The National Schools of Art of Cuba—designed and built between 1961 and 1965 by Ricardo Porro, Vittorio Garatti and Roberto Gottardi—feature similar ideas, in terms of building layout, even though the scope and the degree of sophistication of the project are entirely different. The school spaces are sheltered by shell structures connected through curvy galleries and arcades, and the complex is placed in the middle of an immaculate landscape (Fig. 5.16).

5.3.2 *Binishells and Courtyards*

Similar concepts were explored in the design of the day nursery in Rezzato, Italy. This project consisted of six 15-m Binishells arranged in two groups of three linked by a conventional building volume. In plan, the shell centres are located at the vertices of two equilateral triangles, which are connected via a gallery. Unlike the Cuban

Fig. 5.15 Anna d'Alessandris Pazzi and Angelo Berardi, motel and service station project, November 1970. Internal view of the cupola (DB)

Fig. 5.16 Ricardo Porro, Vittorio Garatti and Roberto Gottardi, National Schools of Art, Havana City, 1961–1965. The School of Ballet (VG)

school project, this arrangement blends the shells into one another, thus generating two internal courtyards. The day nursery playrooms are located underneath the Binishells, while the spaces between the domes are used as recreational and dining areas (Fig. 5.17).

Many of the projects discussed in this chapter—and throughout the entire book—cannot be arranged in a sequence whereby one design represents the chronological evolution of the previous one. Such an approach would not work primarily because several architects and companies were involved in designing and constructing Binishells, and not only Bini.

It is possible to speculate that this day nursery in Rezzato is a morphological evolution of the school in Bologna, whereby the composition and articulation of Binishells were moved one step forward towards the idea of not simply repeating domes but intersecting them. However, this is not the case. Such projects were likely developed independently from each other by different architects and at different times. Therefore, this narrative is purely conceptual and does not purposefully reflect the actual history of facts. The aim here is to highlight some specific projects which—for different reasons—stand out from the hundreds of Binishells that have been designed and built over the past few decades. The idea has been to trace trends and define compositional categories that may be useful to analyse the potential and limits of the Binishell patent, which go beyond the specificities of the individual projects.

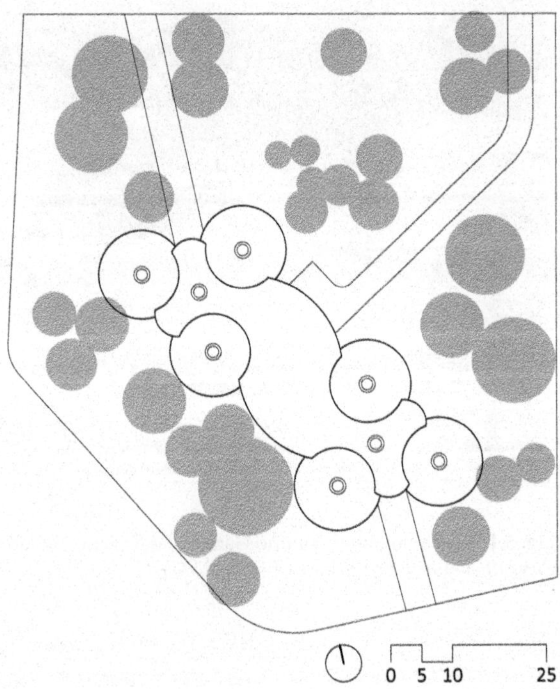

Fig. 5.17 *Binishells spa*, day nursery Binishells, Rezzato, 1969. Site plan (AP)

To summarise, the Torre Cintola holiday resort is presented as the project that encapsulates the basic principle of modularity and repetition of physically independent domes. The Binishell village at the Cappuccini island is then introduced as being—conceptually—one step ahead. Even though the tensile structure that was to link the eight domes was not constructed, the idea of connecting them by adding an external element that also framed the landscape is unique and has not been seen in any other *Binishells spa* projects. Instead, it is more common to find designs, such as the school in Bologna or the day nursery in Rezzato, where additional building volumes were used to connect the shell structures. After this, the subsequent conceptual development in the design of Binishells concerned the physical intersection of concrete domes.

5.4 Intersecting Binishells

Repetition, intersection and arrangement of Binishells in different layouts are complementary design strategies that Bini and other architects used and—in many instances—it is challenging to discuss them separately. It is possible to wonder whether it is logical to exclude the many trimming operations that can be performed on a Binishell structure to create openings from the range of design options. For example, a large cut at the base can change the shape of a shell, thus generating new opportunities for grouping and composing these concrete structures in original ways. Since the focus of this section is on those designs that preserved—entirely or almost entirely—the circularity of the plan, the projects that involved a radical change in the perimeter of the dome are discussed later on.

In a 1969 *Binishells spa* promotional catalogue, where the project in Rezzato was also presented, it is possible to find some diagrammatic design options for other nursery schools that include shell intersections, as well as the compositional strategies seen above (Fig. 5.18). Although these diagrams were conceived to illustrate hypothetical school projects made up of one, two, or three dome structures, they also summarise two recurring design strategies found in many other Binishell projects: two adjacent domes naturally lead to a geometric intersection; three or more domes can instead be juxtaposed to generate an internal courtyard—a shared space that links and expands the functionality of the individual shells. This latter approach was not only used in the day nursery in Rezzato. Several other projects, some of which were built while others remained unbuilt, followed the same layout, as documented in different *Binishells spa* promotional catalogues.

The idea of grouping Binishells and arranging them in circles was also reused in the project for the World's Fair Expo held in Osaka in 1970. A preliminary drawing shows twelve domes, partially intersecting and organised in groups of five, four or three. However, photographs taken at the exposition demonstrate that this concept was eventually realised in a different way: only seven Binishells were built, and these were arranged in two groups of four and three, but without shell intersections. The domes were independent structures connected by narrow arched passageways,

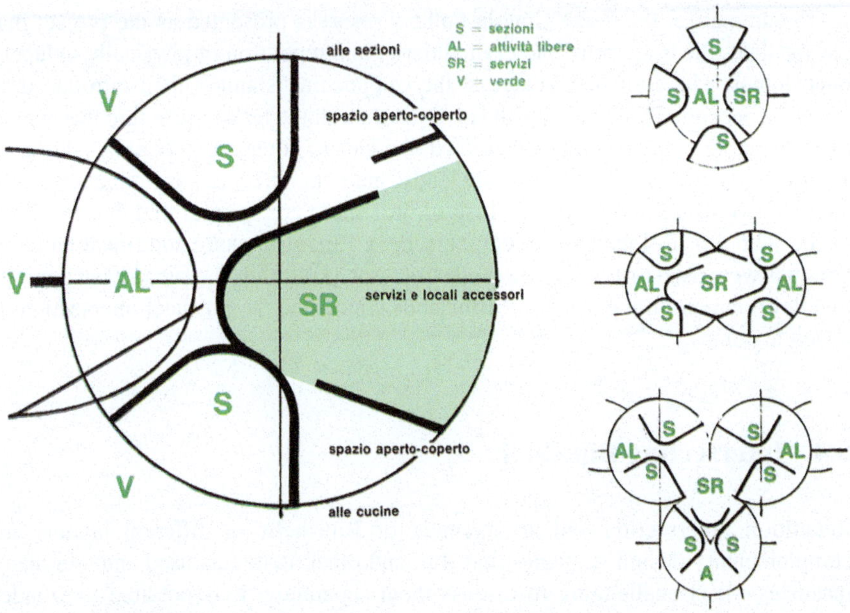

Fig. 5.18 *Binishells spa*, Binishell school schemes from a promotional catalogue, 1969–1970 (DB)

and the outdoor spaces did not seem to respond in any specific way to this spatial composition (Figs. 5.19 and 5.20).

A similar extraneousness to the specificities of the site can be found in many experimental and technological projects of that period, such as La Maison Pneumatique proposed by Jean Aubert, Jean-Paul Jungmann, Auguste Pace, Biagio Pancino and Antoine Stinco for the 1967 Paris Biennale.[1] This conceptual design features dome-like shelters linked via narrow passageways, where the central structure is occupied by the main house rooms and two smaller volumes function as the garage and a recreational area.

Despite the iconicity of the Binishells at the 1970 World's Fair Expo in Osaka, which were an excellent and colourful complement to the pneumatic Fuji Group Pavilion designed by Mamoru Kawaguchi, it is unfortunate the project was not entirely constructed as planned. It would have been a unique example of three Binishells intersecting one another and forming a clover-like shape. Even in the rather ambitious proposal for a large restaurant—designed by the architect Furio Nordio and presented in a 1969 *Binishells spa* promotional catalogue mentioned above—only two of the five intersecting dome structures arranged around a circular courtyard ever meet at a time. Although the catalogue does not include elevation or section drawings of the restaurant, it is clear from the plan that the structures

[1] The project dossier can be found at this link: http://www.urbain-trop-urbain.fr/wp-content/uploads/2011/05/Habiter-pneumatique.pdf. Accessed 24 November 2022.

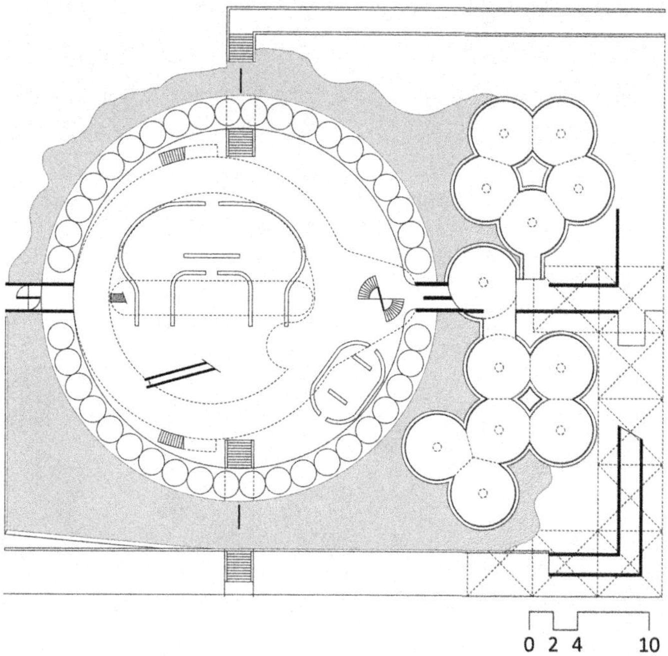

Fig. 5.19 World's Fair Expo, Osaka, 1970. Site plan of the Fuji Group Pavilion and annexed Binishells (AP)

were spaced in such a way that three or more domes never intersected concurrently (Fig. 5.21).

It is common to find projects where two Binishells intersect. For example, the same 1969 promotional catalogue illustrated a design for a 50-m-long sports hall obtained by combining two large domes. This solution creates two seating areas behind the basket poles; it features windows on the longitudinal elevations, while the entrances are placed at both ends of the building—on the short elevations—and are created through a visual lift of the shell base (Fig. 1.29). The Narrabeen North Public School in New South Wales is probably the most famous project that features two intersecting Binishells. It is also the only example remaining in the country. The Space City Shopping Centre in Kallangur, near Brisbane, would have been the second, but—unfortunately—it was not financially successful, and it was abandoned in 1990 and later demolished in 1991. The Binishells at the Narrabeen North Public School were directly designed by Bini himself, while the Space City Shopping Centre was built by *Jennings Industries Ltd* from a project drawn up by a local architect.

The school project in Narrabeen consists of three domes of equal size and shape, two of which intersect. As previously mentioned in Chap. 4, the iconic entrance designed for one of the shells features a sculptural cut that reveals a spherical volume underneath the structure—the so-called pearl of knowledge. The presence of this formal feature explains why that particular view of the building has been published

Fig. 5.20 World's Fair Expo, Osaka, 1970. Aerial photograph of the Fuji Group Pavilion and annexed Binishells (KE)

so much, thus in part overshadowing other valuable aspects of this design. In terms of the repetition and intersection of domes, this project shows a sophisticated volumetric composition, but which is not geometrically complex. It is necessary to pay a visit to the school and walk around the three concrete cupolas to understand the qualities of this small group of Binishells and appreciate their spatial qualities from different angles. They are immersed in a green area with large trees located in the centre of the campus, and they are also surrounded by more conventional buildings arranged in an orthogonal layout, although the open-air theatre is an exception (Fig. 5.22).

The single unconnected Binishell is the first one a visitor would encounter after entering the school campus. It is right there in front of you, behind a few tall trees. The other two intersecting domes are also visible in the background, but their physical connection only becomes apparent when moving further inside, into the middle of the green area. This project can be considered volumetrically sophisticated since the three Binishells conceal each other as a consequence of their positions and due to the shape and placement of the opening cuts. It is difficult to understand whether the shells are blended into a single building or stand as independent units. Moreover, they create different scenarios, depending on the point from where they are seen. For example, from the entrance, the Binishells look like futuristic space shuttles that have landed on earth: they sit on an asphalt surface and are surrounded by a number of more conventional buildings, which make the contrast even more apparent. Further inside the campus—in the green area—the feeling is that the same space shuttles

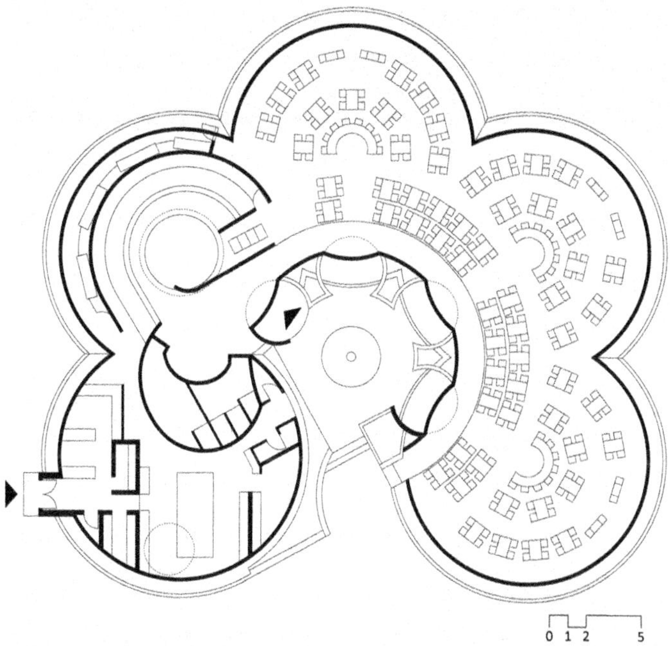

Fig. 5.21 Furio Nordio, Binishell project for a large restaurant, 1969. Plan (AP)

Fig. 5.22 Dante Bini, Narrabeen North Public School, New South Wales, 1974. Site plan (AP)

landed on this piece of land centuries ago, and they are now fully integrated with natural elements that have grown around them.

Although photographs of a building will never replace the experience of a personal visit, they can visually describe the Narrabeen Binishells from three different angles (Figs. 5.23, 5.24, and 5.25). Hopefully, these images can demonstrate that the spatial layout and composition of this project are much more sophisticated than from what it is possible to gather from the plan and images usually published in magazines.

Even though the Narrabeen North Public School and the Space City Shopping Centre are not comparable in size, they are both examples of shell intersections that involve two domes at a time. There were four large Binishells in the Space City Shopping Centre—only three of which physically met—plus three smaller ones. A preliminary sketch of the shopping centre suggests that the shells did not even touch in the initial idea: they were linked by means of a wavy one-story platform, which was used to fill the spaces in between and generate opportunities for the building entrances, circulation spaces and internal courtyards (Fig. 5.26).

This is the project in the entire repertoire of Binishell designs that most explicitly recalls the stereotypical forms of a research base on the moon, which have generally involved a series of artificial dome structures placed in the middle of nowhere and connected by narrow passageways. It is reasonable to assume this analogy was a deliberate design choice, given that the name of the shopping centre was "Space City". It is also very possible that the reference to space symbolised the technological advancement of Bini's construction method, which was so forward-looking for that period to resemble an extra-terrestrial space base (Figs. 5.27 and 5.28).

As seen so far, Bini and a number of other designers experimented with the Binishell system by composing and intersecting concrete cupolas to create large and spatially articulated buildings. The circular base of these structures and—as a

Fig. 5.23 Dante Bini, Narrabeen North Public School, New South Wales, 1974. Photograph from the campus entrance, 9 May 2014 (AP)

Fig. 5.24 Dante Bini, Narrabeen North Public School, New South Wales, 1974. Photograph from the yard, 9 May 2014 (AP)

Fig. 5.25 Dante Bini, Narrabeen North Public School, New South Wales, 1974. Photograph of the intersecting Binishells, 9 May 2014 (AP)

Fig. 5.26 *Jennings Industries Ltd*, Space City Shopping Centre, Kallangur, Queensland, 1978. Preliminary sketch (DB)

Fig. 5.27 *Jennings Industries Ltd*, Space City Shopping Centre, Kallangur, Queensland, 1978. Aerial View (DB)

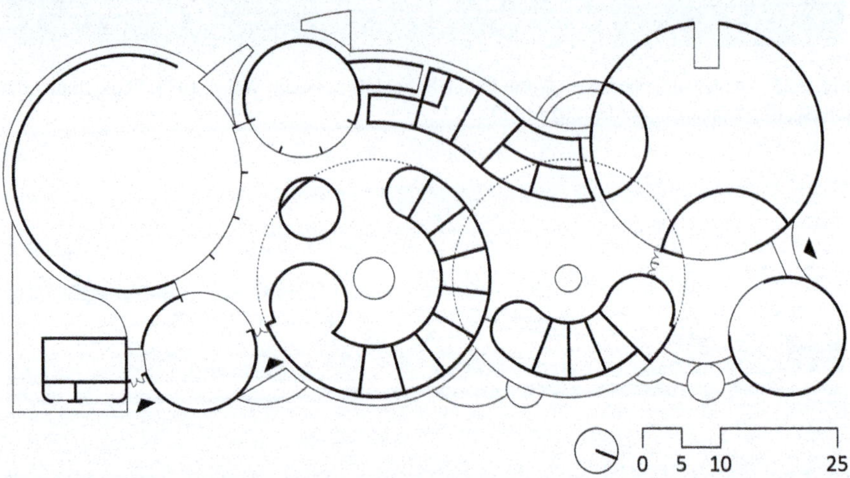

Fig. 5.28 *Jennings Industries Ltd*, Space City Shopping Centre, Kallangur, Queensland, 1978. Ground floor plan (AP)

Fig. 5.29 *Binishells spa*, concept design, 1969. Sketch (DB)

consequence—its final shapes are rarely considered design variables, even though there are a few exceptions. Chapter 3 discussed this point to introduce the Minishell system, which employs an almost square footprint. Bini was well aware of the limitations of designing buildings within circular plans and creating openings on curvy shell surfaces. Therefore, inventing new systems that combined the advantages of the square layout with the speed and economy of construction given by a pneumatic membrane was surely seen as a challenge and a priority at the same time.

Even though the 1969 *Binishells spa* promotional catalogue did not feature the Minishell, it included a sketched perspective of a project consisting of four dome-like structures with an almost square footprint (Fig. 5.29). The advantages of using this solution are immediately apparent, although the design proposal was never actually developed: the insertion and realisation of the openings are simplified, and the articulation of volumes does not require complex geometric intersections.

Changing the shape of the original Binishell system—from variations of the circularity of the plan to manipulations of the shell geometry—does not necessarily imply a departure from the basic ideas that led Bini to develop this new construction technique. It is possible to argue that the original concept will continue to be preserved as long as a pneumatic formwork is used to construct identical concrete shells rapidly and efficiently; but it is also reasonable to contend that a Binishell looks more like a finished building than a component of a larger assembly of concrete structures.

However, there is an exception, which was illustrated in detail in one of the *Binishells spa* promotional catalogues: the modular and prefabricated industrial roof structure already mentioned in Chap. 1 (Figs. 1.19 and 5.30). This construction system is quite unique and even radical among Bini's inventions because—as a concept—it is closer to a vaulted concrete slab than to a large-span shell structure. There is no doubt that a curved slab element, which is directly prefabricated on-site and positioned on a grid of columns—i.e. a "cubic prison"—solves many of the design issues generated by the Binishell system; it also addresses several construction issues related to the repetition and intersection of modular elements.

Fig. 5.30 Dante Bini, Prototype of a modular pneumatic system for industrial applications at the "mushroom field", San Cesario sul Panaro, 1969. Left: the construction of a module on the ground; Right: the built prototype (DB)

5.5 Construction Automation Meets Digital Design

The Binishell is just one of the many construction techniques developed by Bini over the years. In some instances, he created a new method as a direct evolution of a previous one. For example, the Minishell, addresses a specific design issue of the Binishell: the circular plan. In other cases, Bini attempted to broaden the scope for his pneumatic membranes, moving from the mere construction of concrete shells to the erection of pre-assembled steel gridshells, kit homes and even foldable structures.

Nicolò Bini, in a lecture delivered at the Architectural Association in London for the launching of his father's book *Building with Air* (Bini 2014), on 15 October 2015, illustrated the issues he sees in traditional construction methods.[2] He also compared the progress made by the construction industry during the last two centuries with that of another two major industries of comparable size: transportation and communication. In the introductory part of the lecture, Nicolò Bini used the Oxford Dictionary definition of the term "architecture" to make a point, namely that "doing architecture does not only consist of designing". He also said that this definition is 50% of the equation, and the remaining 50% should be dedicated to constructing. To demonstrate how far we are from this ideal balance, Nicolò Bini highlighted that the way we are currently building does not differ so much from the innovative methods developed years ago for such key projects as the 1851 Crystal Palace in London and the 1931 Empire State Building in New York. According to him, most of the recent innovations have taken place during the design phase, while the construction industry has remained somewhat stagnant.

Even though there is a certain degree of simplification, the scenario illustrated in this lecture is embraceable, and it was ideal for introducing the work and legacy of the renewed US *Binishells Inc.* company, which mainly focuses on developing automated and energy-efficient construction techniques. However, the developments that have taken place in digital design in the last few decades have undoubtably opened

[2] This lecture can be watched on YouTube at: https://youtu.be/vl6Vvf1J1og. Accessed 15 October 2022.

up new design opportunities for architects and engineers, including Nicolò Bini. They can now conceive free forms directly in a CAD environment using parametric curves and surfaces—the Non-Uniform Rational B-Splines (NURBS).[3] Such curves and surfaces are easy to manipulate through control points, and they are currently implemented in commonly used 2D drafting and 3D modelling software, such as Rhinoceros®. Nicolò Bini can therefore simulate a pneumatic form-finding method digitally, and he can calculate complex structural forms through a Finite Element Analysis (FEA). Such design freedom can be expected to make the concepts of modularity, repetition and geometric intersection obsolete. However, the reality is never so straightforward, and this statement is only valid for some of the most recent Binishell projects but remains invalid for others.

The villa in Malibu designed for the actor Robert Downey Jr is an example of the former. This project, which is analysed in detail in Chap. 6, features a single sinuous concrete shell roof that does not include any modular elements. It almost looks as if this structure was modelled by dropping a sheet over the rooms it covers, where openings were cut at some points along the edges to provide access, light and air to the dwelling. The membrane used to erect this modern Binishell, called System A, was created ad-hoc for this project, thus eliminating the possibility of reusing it in future designs with different formal features (Fig. 1.32).

The layout of the house certainly shows an affinity with the planimetric organisation of the school in Rezzato, or even with the Space City Shopping Centre, if the obvious differences in scale are excluded. However, the shell shape of the villa in Malibu does not reflect—or communicate—the position of the circulation spaces as clearly as the older projects did. Instead of having a direct correspondence between living spaces, which are generally placed underneath the shells, and the circulation spaces, which are generated by adding other building volumes, such as galleries and arcades, this complex structure blends everything under a single sinuous roof.

Nicolò Bini used System A in a few other conceptual design proposals, ranging from residential applications to holiday resorts. Images of these unbuilt projects have been disseminated online. Although the compositional rules of repetition and geometric intersection introduced above do not apply to many of Nicolò Bini's projects, several of his designs recreate the spatial qualities explored in the past by his father and other architects who used the original Binishell system. For example, as already pointed out in Chap. 1, an experimental house project (Fig. 5.31), which was published on the cover of the Italian book *A cavallo di un soffio d'aria. L'Architettura Autoformante* (Bini 2009), recalls the large self-service restaurant designed by Nordio, with five domes arranged in a circle and the presence of a central courtyard (Fig. 5.22); a similar planimetric layout can also be found in a wellness centre project in the Azores (Fig. 5.32). Another conceptual project for a

[3] NURBS (Non-Uniform Rational B-Splines) curves and surfaces can be considered the current standard for the definition and modification of curves and parametric surfaces in CAD software for architecture and industrial design, such as in Rhinoceros®. Originally developed for the automotive sector, NURBS curves and surfaces allow a designer to create and manipulate free-form geometries through the use of control points (Piegl and Tiller 1997).

holiday resort features groups of Binishells that are well integrated into the landscape (Fig. 5.33): they recall a lunar base and—to a certain extent—resemble those designed by Dante Bini for the Narrabeen North Public School (Figs. 5.24, 5.25, and 5.26).

Apart from the projects developed with System A, Nicolò Bini has also promoted three other systems—called B, D and E—all of which preserve the modularity of his father's invention. System B is a modern version of the 1979 Minishell, a structure with a square footprint that works well as a stand-alone unit, while Systems D and E are more recent designs featuring anticlastic curvature. They also work as independent structures and can be used to build terraced houses or holiday resorts.

In the last few decades, the concept of construction efficiency has evolved to such an extent that the first Binishell patents can no longer be considered innovative for our current needs and normative frameworks. The term "efficiency" still means building rapidly and using standard or modular elements in the built environment, but it can also refer to environmental sustainability criteria or waste reduction strategies. Therefore, it makes little sense to compare Nicolò Bini's projects with those of his father, apart from when referring to how construction automation techniques for shell structures have informed these designs.

The original Binishell system acted simultaneously as a form-finding and construction method. The design process was somewhat constrained because the boundary conditions of the system—for example, the shell diameter and membrane shape—defined the structural form and dictated the compositional rules of the system,

Fig. 5.31 Nicolò Bini, experimental house project, 2009. Bird's eye view (NB)

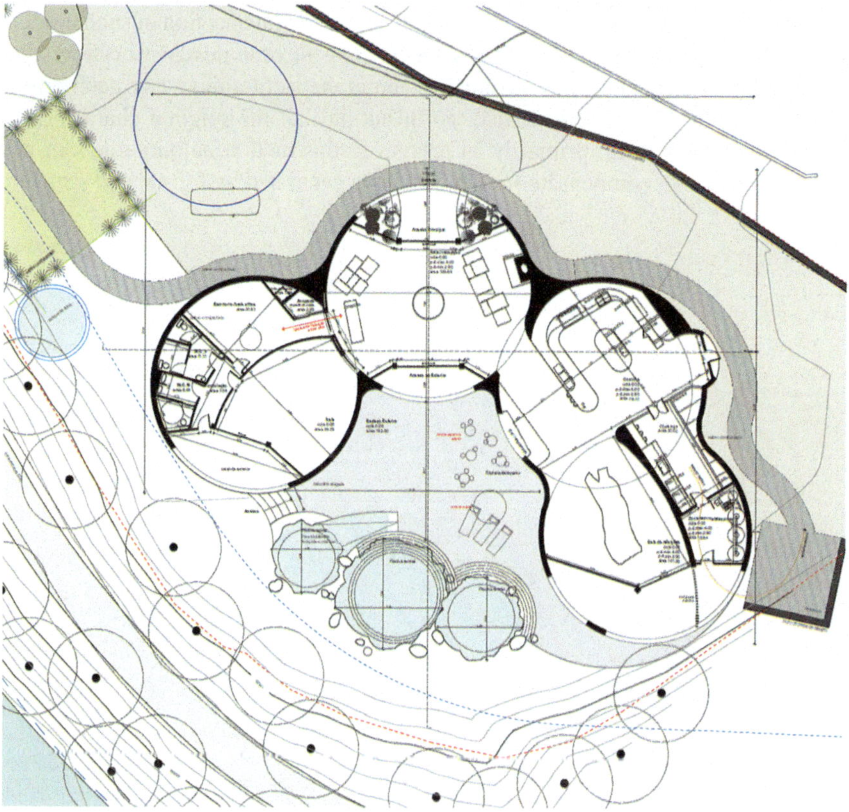

Fig. 5.32 Nicolò Bini, wellness centre project, Azores, November 2022. Site plan (NB)

Fig. 5.33 Nicolò Bini, holiday resort project in Greece, 2014. Bird's eye view (NB)

including the opportunities for repetition and geometric intersection of such modular domes. Nicolò Bini can now follow a much more flexible design workflow since he can adapt the shape of the pneumatic membrane to fit almost any geometry he wishes to build. However, it is also worth noting that the original characteristics of the Binishell system, primarily in terms of construction automation, can now become a means to compensate for the problems generated by formal and structural complexity.

Chapter 6
Dante Bini's Legacy Beyond the Cupola

6.1 Construction Innovation in a World of Free-Form Architecture

Designing with form-resistant structural typologies, a category into which concrete Binishells fall, implies generating non-orthogonal spaces. Such spaces, inevitably, go way beyond the more common verticality of columns and walls or the horizontality of floor and roof systems. Investigating the formal evolution and the composition of such spaces is a highly topical architectural theme, especially now that computational design, Finite Element Analysis solvers and digital fabrication techniques do not create any theoretical limits for designers. The contemporary panorama of this set of design research topics is much more complex and articulated than it was back in 1964, in the early days of the Binishell patent. In fact, it is much more challenging to find patterns in the so-called free-form architecture of the last thirty years than to categorise the vast production of shell and tensile structures of the first half of the twentieth century. Traditionally, shell and tensile structures were conceived using either geometric forms or one of the few form-finding techniques that were available at the time, with the overall aim of achieving an optimal structural form or a certain degree of construction rationality. The development of digital tools has drastically changed this situation: today's architects are constantly pushing the boundaries of the more conventional design processes, techniques and construction methods to explore new ways of conceiving and building architectural spaces. In such a complex scenario, it is still critical to analyse how the constructive response to formal and functional desires is calibrated, case by case, with the realities of the local construction industry. Purely technical aspects, such as the preparation of the formworks, the positioning of the reinforcements or even the techniques used for casting the concrete, can still significantly affect the tectonic quality of the final building, regardless of the technological advancements in digital design and fabrication.

The number of research projects dedicated to developing and testing new construction and fabrication techniques has increased in the past couple of decades. Some

A. Pugnale and A. Bologna, *Architecture Beyond the Cupola*,
Mathematics and the Built Environment 7,
https://doi.org/10.1007/978-3-031-26735-2_6

researchers in the field of concrete construction have tested fabric formworks to build concrete components that are formally complex. For example, Alan Chandler and Remo Pedreschi—two academics working in the UK—have focused on prefabricated columns and wall components, prioritising texture and surface quality in the prototypes realised by their teams (West 2017: 161, 189). The Block Research Group (BRG) at ETH Zürich has worked on stay-in-place knitted formworks for the construction of thin concrete shells. In collaboration with Zaha Hadid Architects, BRG built the 2018–2019KnitCandela prototype, a shell that features an anticlastic double curvature that pays homage to Félix Candela and demonstrates the feasibility of using weft knitted fabrics as lost formworks (Popescu et al. 2021). The Los Angeles-based architectural firm Form Found Design has instead experimented with robotically controlled Y-shaped fabric formworks to build free-form concrete gridshells and spatial structures, such as the 2017 MARS Pavilion and the 2018 Cytocast prototype (Stelsel 2018). Other researchers and companies have developed adaptive moulds to facilitate the construction of double-curved façades. For example, the Danish company *Adapa*, which was created as an Aalborg University spin-off in 2010 by two young graduates, is now a well-established business that works on the design and construction of complex cladding elements in various countries (Raun et al. 2011).

However, it is possible to state that the employment of pneumatic formworks with the aim of speeding up, simplifying or even automating the construction process of concrete shells is very rare, even in those cases in which the design presents formal qualities that would seem to suggest their use. Two recent projects built in very different countries and circumstances—in Switzerland and in China—provide evidence that relatively conventional construction techniques, based on wooden formworks, are still adopted nowadays for designs that feature a high degree of geometric complexity.

The first of these projects is the Oval Centre in Chiasso, in the Ticino Canton, in Switzerland, which was completed in 2011 according to the design of the architect Elio Ostinelli and the engineers Aurelio Muttoni, Franco Lurati and Miguel Fernández Ruiz. The second is the UCCA Dune Art Museum in Qinhuangdao, China, which was finished in 2018 by OPEN Architecture and was designed in collaboration with the Design Institute CABR Technology Co.

The scale and purpose of these two projects are radically different, but both required complicated digital modelling and fabrication techniques to prepare the formworks that gave shape to the double-curved concrete structure surfaces. Although the use of pneumatic membranes would surely not have resolved all the problems connected to the waste and recycling of the traditional wooden formworks, and the presence of an even more skilled labour force would perhaps have been required, it is nonetheless necessary to discuss these two projects in this chapter dedicated to Bini's legacy. The aim of this discussion is in fact to evaluate his inventions within the broader panorama of contemporary construction processes and techniques, particularly those employed to erect concrete structures of high formal complexity.

As suggested by its name, the Oval Centre in Chiasso is basically a large elliptical shell whose main axes are 92.8 m long, 51.5 m wide and 27 m high. The thickness of the shell varies from a minimum of 10 cm to a maximum of 12 cm, and it contains four reinforcement layers. The central part is 12 cm thick and is prestressed. Moreover, the concrete in this zone was further reinforced with steel fibres. The wooden formwork was designed to be used—and then removed—in two stages, starting from the lower part of the shell (Figs. 6.1, 6.2, and 6.3), which was obtained by spraying concrete directly onto the formwork. The upper part, where the slope is less than 15 degrees, was obtained by casting the concrete in place. It had a complicated centring made of glulam beams, placed at a centre distance of about 2 m, to which secondary structural elements were attached to support the plywood panels of the formwork (Muttoni et al. 2013). Because of its shape, this upper part could have been substituted, theoretically speaking, by a large pneumatic membrane.

Li Hu and Huang Wenjing, founders of the OPEN Architecture studio in Beijing, adopted a similar construction choice for the building of the UCCA Dune Art Museum in Qinhuangdao. The design of this museum is the result of a distinctive Chinese design culture, which is able to liberate, with nonchalance, the generation of structural forms from basic principles of geometrical logic, constructive efficiency and material properties. With this approach, a structure is not generated as an optimal form but rather as a purely spatial system, which may derive from, for example, just the perceptive and sensory effects desired by the architect. Constructed among the dunes of a beach that faces onto the Yellow Sea, the museum is made up of a sequence of underground rooms of an amorphous geometry, which, at a first glance, appear to be shell shapes carved out of the rock. The structural design of the museum was

Fig. 6.1 Elio Ostinelli, Aurelio Muttoni, Franco Lurati and Miguel Fernández Ruiz, Oval Centre, Chiasso, Ticino Canton, 2011. View of the wooden formwork (LMP)

Fig. 6.2 Elio Ostinelli, Aurelio Muttoni, Franco Lurati and Miguel Fernández Ruiz, Oval Centre, Chiasso, Ticino Canton, 2011. A photograph taken during the concrete spraying stage (LMP)

Fig. 6.3 Elio Ostinelli, Aurelio Muttoni, Franco Lurati and Miguel Fernández Ruiz, Oval Centre, Chiasso, Ticino Canton, 2011. Internal view of the raw concrete shell (SM, LMP)

developed through a constant dialogue between the architect and the engineers, who were involved in a synergetic job of iterative revision and progressive adaptation of the structural requirements to the architectural choices, and vice versa. The process involved alternating analogue tools, such as physical models (Fig. 6.4), with digital software for parametric modelling and Finite Element Analysis. The final structure presents a variable cross-section that ranges between 40 cm at the base to a minimum of 25–30 cm at the top. This choice results from the significant load of soil, water and vegetation that the reinforced concrete shells have to bear.

The free-form geometry of each underground room was constructed by resorting to a wooden formwork in this project, even though a whimsical hypothesis of casting concrete onto a mass of ice has initially been advanced by the architect Li Hu (Fig. 6.5). This was the cheapest solution since local workers from Qinhuangdao were employed, and many of them had had previous experience in shipyards in the construction of wooden boats. Elastic material was used in combination with small timber boards to prepare the formworks, especially for the excessively curved zones. It is worth noting that the architects willingly agreed to maintain the uneven and raw surface texture left by the formworks on the concrete surfaces, thus allowing the construction technique to be clearly expressed in the final building and to act as an ornamental system of the museum structure (Fig. 6.6). This is an honest construction

Fig. 6.4 OPEN Architecture, UCCA Dune Art Museum, Qinhuangdao, 2018. 3D-printed models of the structure (OA)

Fig. 6.5 OPEN Architecture, UCCA Dune Art Museum, Qinhuangdao, 2018. View of the wooden formwork (OA)

that was further enhanced by the use of white paint, the idea behind which was to spread the natural light in the best way possible and, at the same time, to emphasise the complex shape of the internal spaces (Bologna 2020).

Resorting to traditional formworks, whether made up of wooden boards or using engineered steel components, offers various guarantees to the designer and the builder, compared to the use of pneumatic membranes. These guarantees are, above all, dictated by the employment of consolidated techniques and practices, which have been used throughout the world to cast reinforced concrete structures for over a century. Furthermore, resorting to traditional formworks rather than a pneumatic membrane ensures an unconditioned stability of the structure being built. Using air, and therefore inflation pumps to manage the air pressure, could easily lead to having formworks that continually change, even though slightly, in shape and size.

In spite of Bini's endeavours over almost half a century to persuade clients, designers and builders about the value of his patents and systems, as much for the conception of quality architectural spaces as for the practicality and cost-effectiveness of the techniques, the resort to pneumatic formworks has never managed to undermine the more traditional methods used for building reinforced concrete shells. In fact, the hundreds of domes created with the Binishell system continue to be considered as the result of astute entrepreneurship and open-mindedness to experimentation more than a procedure that has become consolidated over the years, even outside the construction sites that Bini and his companies managed directly.

Even the September 2005 American Concrete Institute (ACI) report, entitled *Construction of Concrete Shells Using Inflated Forms,* presented Bini's invention

Fig. 6.6 OPEN Architecture, UCCA Dune Art Museum, Qinhuangdao, 2018. Internal view of the raw concrete structure (OA)

as a singular and unique construction method, and therefore concentrated on other applications of the pneumatic membrane in the concrete construction industry. The report included techniques that have been fine-tuned, starting from the invention that was already patented back in 1941 by Wallace Neff, in which the membrane is inflated before positioning the reinforcement rods and casting the concrete, and it therefore functions as a surrogate for a wooden formwork. The ACI report stated that the: "inflated-form, thin-wall shotcrete construction has become one of the most common and widely used methods in the construction of domes", so much so that "The Monolithic Dome Institute estimates over 2,000 thin shells have been built over the last 30 years using the fabric form method, whereas those built with conventional forming methods are few in number". Nevertheless, the same document also explicitly mentioned that "until recently, only a few contractors have possessed the skills and equipment necessary to undertake this type of construction", but "as architects and engineers are becoming aware of the advantages of this inflated form method and its use increases, industry design and construction standards are needed" (American Concrete Institute 2005: 334.3R-3).

Nicolò Bini is one of the designers who, through his experimental projects, has for some time been demonstrating the validity of the pneumatic formwork for the construction of reinforced concrete shells. His work is also essential to evaluate up to what point the evolution and revision of the technique invented by his father Dante can respond to the necessities and requirements of the present construction industry.

The Oval Centre in Chiasso and the UCCA Dune Art Museum in Qinhuangdao are clear examples of how contemporary architects are continuing to explore the relationships between architectural and structural form and construction techniques, especially for reinforced concrete roofing. However, the limits of the original Binishell system, as conceived by Dante Bini back in the 1960s, are amplified when designers force the architectural form beyond the conventional dome geometries or even the funicular shapes derived from traditional form-finding techniques, such as the reverse hanging method.

Therefore, the current application of the original Binishell system raises certain practical issues: the evolving building codes, although in no way affecting the validity of the patent from the construction rationality point of view, question the validity of such a construction method because of the new requirements of the contemporary construction industry. For example, how does the Binishell system respond to the current energy and seismic regulations? Up to what point can the modular pneumatic formwork be applied to the current formal explorations and architectural languages generated through free-form NURBS surfaces? These are key design aspects that cannot but cast doubt on the entire conception of Dante Bini's systems. Nevertheless, in a proactive way, it is also necessary to question ourselves about what the points of strength of these systems are in order to understand to what extent a further development of such knowledge and technology could respond to today's problems. For example, the optimisation of the energy necessary for the construction of a building becomes a crucial parameter for the overall evaluation of the degree of sustainability—both environmental and economic—of an architectural work.

6.2 Nicolò Bini's BLOBs

The villa designed by Nicolò Bini between 2012 and 2013, and then built between 2015 and 2017, takes on a particularly relevant meaning, when considering the aforementioned grounds. This villa was constructed on a lot between Morning View Drive and Ebbtide Way in Malibu, California, by the US *Binishells Inc.* company, of which Nicolò is Managing Director and President, while his father Dante is in charge of Research and Development. The relevance of this project concerns not only the architectural character of the building, which results from a pneumatic form-finding process performed from a free-form plan boundary, but also the development of a construction system which could be compatible with the site practices currently used in the Los Angeles area, in California.

Like the concrete cupola designed and built for Michelangelo Antonioni in Sardinia, this Binishell in Malibu was commissioned by a celebrity: the actor and film producer Robert Downey Jr. It was therefore a singular project realised for a specific site and developed in collaboration with a private client. It can also be considered as the last link of the evolutionary chain of the Binishell patent: a sort of great-grandchild of the first experimental dome built in Crespellano in 1964.

Nicolò completed his bachelor's degree and then his master's degree in architecture at the University of California, Berkeley, where he developed an interest in low-cost housing systems for developing countries. He then perfected his studies at the University of London and at the Sapienza University of Rome and worked, between the 1980s and 1990s, in various architectural firms in London, Rome, Florence and San Francisco, before founding his own firm, Line Architecture, in the US in 2005. Nicolò also appears beside his father, as both a designer and supervisor of keynote presentations which are presented at conferences throughout the world. His has been a transversal training, which has included combining solid theoretical foundations with pragmatic experience that he has gained while following the designing and construction of Binishells. Over the years, Nicolò has developed a design workflow which, like that of his father, has led him to explore the architecture of concrete shell structures through the development of non-traditional construction techniques by means of 1:1 scale prototypes. However, Nicolò has also shown a particular sensitivity to sculpturing architectural spaces, which does not seem to be entirely driven and constrained by structural form finding or construction technology: his design approach primarily embraces building and rationality principles to inform architectural responses that can satisfy the most recent structural and environmental performative requirements. "My father's work shows us how a simple, but incredible radical concept can invert many of the contemporary notions of construction and design", wrote Nicolò, who revealed that "the most powerful lesson I have derived from a lifetime of observing my father's work and way of thinking, is the elegant and courageous example he has set for us to question and re-invent the way we use our resources to help us design our environment and to build our communities more efficiently" (2014: 146).

Nicolò relaunched his father's company as *Binishells Inc.* in the US in 2010, mainly working on the improvement and fine-tuning of the technologies used for erecting thin concrete shells erected through the use of a pneumatic formwork. In 2016, Nicolò explained how his revised and new construction systems offer significant advantages over the older Binishells and the related patents, and over other air-based technologies: "These advantages include reduced construction time, cost, complexity and environmental impact, with enhanced architectural desirability and flexibility. In regards to the latter, our systems enable square, rectangular and asymmetrical plans as well as multi-story typologies and are code compliant internationally" (Bini 2016). Nicolò's objective is to make his company's construction systems competitive on the current residential market, mainly by promoting them in terms of energy and economic efficiency: "our new systems are more affordable, safer and more environmental than comparable technologies resulting in code compliant structures. In synthesis our Systems are more affordable because they require less labour and materials, safer because they are more efficient structurally, and more environmental because they provide thermal bridge-free building envelopes" (Bini 2016). However, it is clear, from his designs, that Nicolò's agenda also has the aim of revisiting the formal expressions and the architectural languages of the projects he has developed using his Binishell systems, most likely being inspired by the many recent free-form architectural works that have been created.

The villa in Malibu, if analysed from a purely formal point of view, from its amoebic shape to the quality of the double-curved concrete surfaces that define the internal spaces, can be fully regarded as part of a trend of *non-standard* projects which, in the last few decades, have gained great media success thanks to the spread of appealing images via international exhibitions, as well as the social media and the web. It is sufficient to think about the extraordinary success the MAD Architects firm is having with the Wormhole Library project, located inside Century Park in Haikou, China, which was completed at the end of 2021. The sinuous shapes of the building, just like the amoebic voids that characterise the internal spaces, were created in white concrete and cast from traditional, although geometrically complex formworks, cut by means of Computer Numerical Control (CNC) routers and completed with 3D-printed components to achieve surfaces as smooth and continuous as possible (Figs. 6.7 and 6.8).

It is important to underline that Nicolò's creative impulses in the conception of single-family houses, whose spaces are, in this case, generated using concrete shell structures are, in many ways, comparable with the research which, already back in the 1950s, led to the development of some well-known experimental projects, such as John M. Johansen's Spray Concrete House #2, built in 1955 (Fig. 6.9), or Frederick J. Kiesler's Endless House designed in 1958 (Fig. 6.10). The latter, although it was never actually constructed, was able to gain great popularity thanks to the suggestive idea of creating a domestic space inside a concrete shell that seems to be floating in air, but also in part because eight different models that had been

Fig. 6.7 MAD Architects, Wormhole Library, Haikou, 2021. View of the wooden formwork (A, MA)

Fig. 6.8 MAD Architects, Wormhole Library, Haikou, 2021. View of the finished building (AE, MA)

made from 1949 onwards were exhibited at the MoMA in 1960, on occasion of the *Visionary Architecture* exhibition, curated by Arthur Drexler. This project was a sort of architectural manifesto, which "develops the surface of the building as a twisting, continuously curved ribbon wrapped around itself" (Drexler 1960: 1). The "treatment of the wall surface would produce a building more like sculpture than architecture" (Drexler 1960: 8) and recalls the *new sensualism*: a current, as defined by Thomas H. Creighton in *Progressive Architecture* at the end of the 1950s, that is capable of imposing itself as the antithesis of the box architecture presented by *International Style* (Rosa 2001: 8). Some lesser-known examples, designed on the basis of the same assumptions, are surely relevant in this discussion because they were not just experimental projects—they were actually constructed. Mention can be made of the concrete shell houses built in France at the beginning of the 1960s, designed by Pascal Häusermann,[1] or those by Jacques Couëlle,[2] whose spatial qualities can surprisingly be considered topical, especially when compared with those of Nicolò's

[1] This project refers to numerous buildings, such as the Egg-Shaped House, designed by Pascal Häusermann (with Bruno Camoletti and Eric Hoechel) in Pougny, Ain, France, built in 1962, or the Hotel Thierry building in Raon-Stem, France, constructed in 1967.

[2] As an example, it is worth mentioning his house-sculpture in Mouans-Sartoux, France, built in 1962.

villa in Malibu and the spaces of the UCCA Dune Art Museum or the Wormhole Library.

The small domestic buildings of Johansen, Kiesler, Häusermann and Couëlle highlight a manneristic approach to the exploration of architectural form, which has undoubtably had, over the last seventy years, a surprising temporal, as well as constructive-technological continuity and a global diffusion, as developed through experimental projects and a theoretical systemisation of the architectural design discipline. This formal research also led to the term *non-standard architecture* being coined, following the eponymous exhibition, held between December 2003 and March 2004, at the Centre Pompidou in Paris, which aspired to collect a vast and heterogeneous set of experimentations in the field of digital design and fabrication. As Zeynep Mennan pointed out: "the non standard inscribes itself within the realm of contemporary architectural experimentations making extensive use of recent computational design technologies and its formal catalogue is marked by highly complex

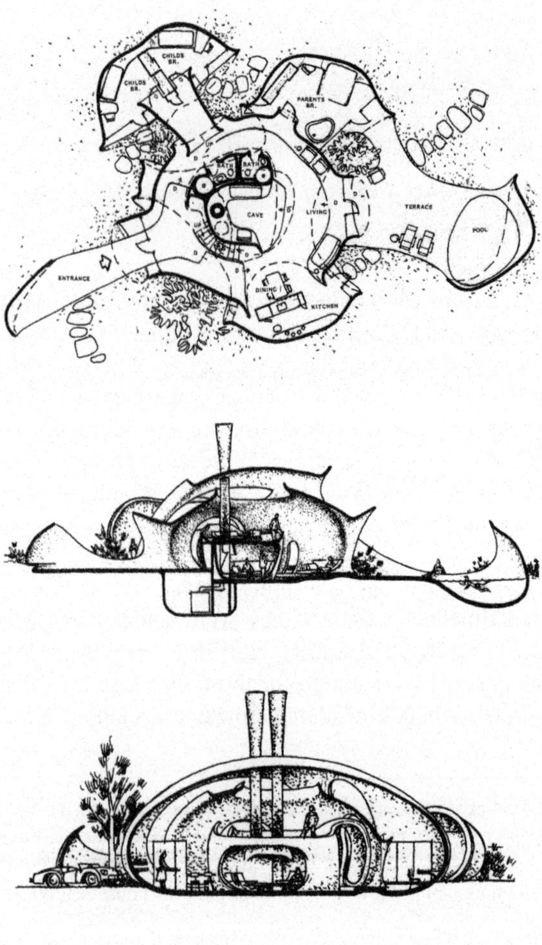

Fig. 6.9 John M. Johansen, Spray Concrete House #2, 1955. Plan and sections (JMJ)

Fig. 6.10 Frederick J. Kiesler, Endless House, 1958. Photograph of the model (MoMA)

dynamic forms that indicate a revival of the organic tradition" (2008: 171). The *Non Standard Architectures* exhibition presented a variety of approaches, which ranged from the purely virtual explorations of Marcos Novak—unconstrained by tectonic principles or construction techniques—to works that linked free-form designs to new digital fabrication methods, in a continuous process that is generally referred to as "file-to-factory".

Nicolò revamped the *Binishells Inc.* company in this complex period, in which design processes and techniques are in a state of continuous change as a result of the rapid evolution of digital design and fabrication. Freeing himself from the most rigid form-finding and construction process of the original Binishell system, Nicolò has aligned his work with a design philosophy whose expressive forms are based on the theoretical assumptions that developed from the positions adopted by Greg Lynn towards the end of the 1990s. His villa in Malibu may be regarded as part of the family of the so-called Binary Large Object (BLOB) architecture because of the sinuosity of its shapes. From the desire to utilise and update the construction techniques initially developed by his father, Nicolò explores and defines his own approach to the design of Binishells, thereby becoming another advocate of an architectural mannerism of our times. Obviously, this means a design context that is generationally very different from that in which Dante worked.

Nicolò defined his formal explorations, which are based on technological improvements and evolutions of the Binishell system, with the term *bio-metricism*, which he describes as a "contemporary design philosophy based on a return to the tradition of looking to nature as an aesthetic and functional inspiration" (Bini 2016). Thus, his new Binishell "represents a radical, yet simple construction concept whereby natural principles and forces may be harnessed to build sinuous, curvilinear structures easily, quickly and environmentally" (Bini 2016). This is a design philosophy that merges the BLOBs of Lynn with the "architecture of necessity" of Frei Otto (Otto and Rasch 1995).

Nicolò, starting from technological advancements and innovations in construction, has the explicit ambition of giving life to a distinct architectural formalism: "by introducing this new aesthetic via recognised taste-makers, it may be possible for Binishells to take hold at all levels of the market" (Bini 2016). It is an approach that is aimed at attracting a sophisticated type of clientele for bespoke projects, but which does not negate the potential for standardisation and serial production that has guided the development of various Binishell systems from the 1960s, as Dante's Villa Antonioni has already demonstrated.

6.3 A Villa in Malibu

The villa that Nicolò designed for Robert Downey Jr in Malibu was probably inspired by the fascinating formalisms and spatial qualities of the 2010 Teshima Art Museum by Ryue Nishizawa and Rei Nato. It has a footprint of about 558 square metres; it stands as a sculptural object in the middle of a small private park, where it is deliberately hidden from the public access road and opens up towards the beach in the southwest direction.

The form-finding-to-construction process for this Binishell house certainly followed a non-conventional path: it is possible to recognise—at the same time—pure creativity derived from an empirical approach to design and a degree of construction rationality related to functional requirements, as well as an understanding of the structural behaviour of concrete shell structures. It is also worth noting that Nicolò demonstrated a great deal of initiative and courage in leading this project, from the sketch design phase to completion of the building. In fact, it is important to remember that he did not conceive this house in the Italian post-war context of the 1960s, in which his father operated: that was a flourishing period for the concrete construction industry, characterised by design experimentation, technical advancements, and the low cost of labour. The final result is a sinuous thin concrete shell that shows remarkable coherence between architectural and structural form, as well as between the form-finding method and the employed construction technique (Fig. 6.11).

It is worth underlining that Nicolò began designing the villa in Malibu from the plan: he studied the shape, size, arrangement and flow of the rooms in the house using a relatively conventional design tool, but he also envisaged and took into account—at the same time—the tridimensionality of the roof structure that would have covered such functions. In doing so, he primarily used curves to generate irregular and non-orthogonal spaces that then became the distinctive formal feature of the

Fig. 6.11 Nicolò Bini, villa in Malibu, 2012–2017. Aerial photograph of the finished building (NB)

house (Figs. 6.12 and 6.13). When drawing the plan, Nicolò defined the boundary conditions that were necessary to perform a pneumatic structural form finding, as originally conceived by Heinz Isler in 1954 under the name "inflated hill" method (Isler 1994) and then tested several times by Frei Otto (Vrachliotis et al. 2017).

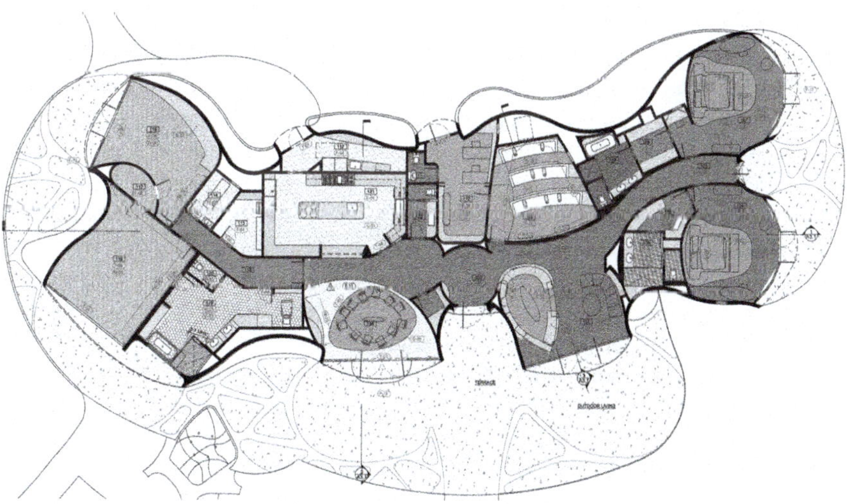

Fig. 6.12 Nicolò Bini, villa in Malibu, 2012–2017. Plan (NB)

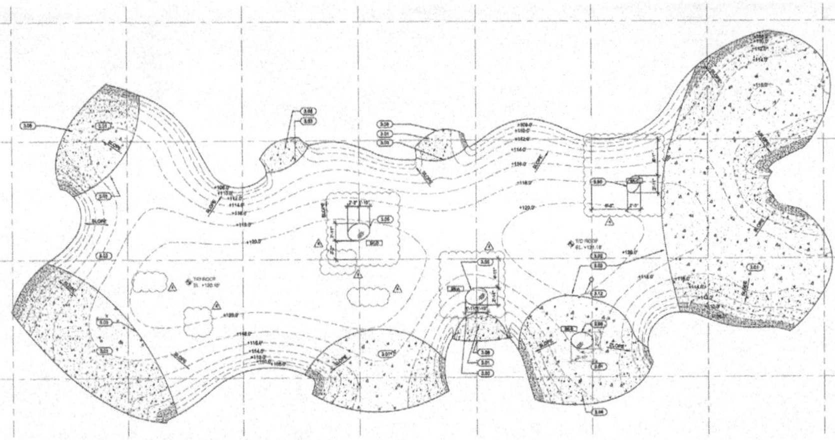

Fig. 6.13 Nicolò Bini, villa in Malibu, 2012–2017. Roof and pneumatic membrane plan (NB)

Pneumatic form finding is a particularly fascinating structural design method, but it is also a restrictive one for a designer, because the only main architecturally relevant variable of the process is the shape of the shell footprint: the perimeter of the building practically defines a specific tridimensional structural form that results from the inflation of a membrane that is constrained along the edge of such a boundary (Figs. 6.14 and 6.15). It is in fact an iterative process of trial and error, in which the designer simultaneously develops the plans and the sections, while trying to imagine the spatial qualities of the interiors by working on a bidimensional plan drawing.

Although Nicolò's design process is based on a pneumatic form-finding method that is similar to that of his father, the two architects operate on the basis of radically different principles. On the one hand, Dante Bini conceived his projects as being bound by the rules of modularity and constructive efficiency, and he therefore used a well-defined set of compositional rules for the design of his circular Binishells, but also of his other systems with a square, rectangular or hexagonal plan. On the other hand, for about twelve years, Nicolò has been developing his own construction systems, some of which can generate BLOB structures that can be adapted to planimetric configurations that are not necessarily modular or repeatable. Nicolò has freed himself from the shape of the cupola, but also from the main advantage of the old Binishell system, which is that of being able to place the reinforcement and pour the concrete onto the ground, onto a flat surface, before inflation. However, this does not seem to be a problem, given that the old system no longer complies with the current building codes, and cannot therefore be used as it was originally intended.

Nicolò designed the architectural and structural form of the villa in Malibu by resorting to the Rhinoceros® parametric modelling software. Starting from a curvy and sinuous plan perimeter, he simulated the inflation of a pneumatic membrane by means of a physics engine acting on a triangulated mesh. An architect can adjust and perfect the overall shape of a structure in the digital domain by acting on a limited number of design variables, such as the control points of the parametric curves

Fig. 6.14 Frei Otto, an experimental pneumatic form-finding model, August 1960 (FO)

Fig. 6.15 Frei Otto, an experimental pneumatic form-finding model, August 1960 (FO)

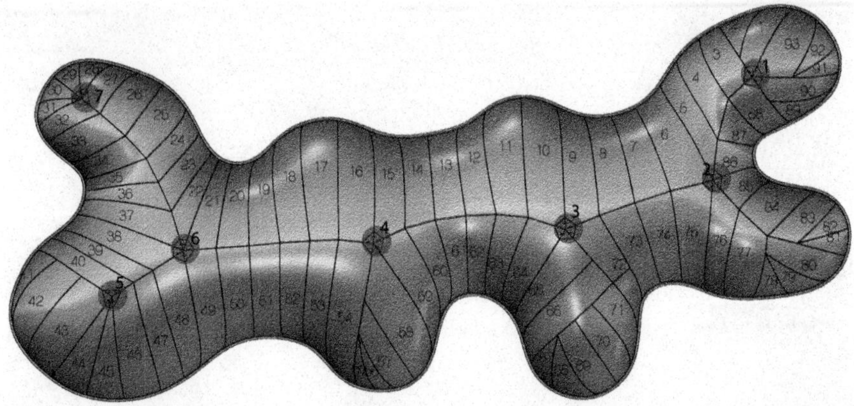

Fig. 6.16 Nicolò Bini, villa in Malibu, 2012–2017. Pneumatic formwork discretisation, plan (NB)

that define the perimeter of the house. The final mesh geometry was then exported into a FEA software for the structural analysis and dimensioning of the concrete shell. A NURBS surface of the structure was also constructed to fabricate the PVC coated polyester fabric pneumatic formwork. This surface was discretised into 93 narrow strips that could be developed, i.e. unrolled onto a plane; 7 singular points that remained on the surface were patched using irregular polygonal geometries (Fig. 6.16). These fabric sheets were then welded. This entire process was carried out directly by the manufacturer of the membrane.

Nicolò has also made physical models, by resorting to vacuum-formed clear plastic and 3D printers, to develop this kind of residential projects, which are based on complex curvilinear footprints. However, these models cannot be considered actual form-finding tools, even though there are some exceptions that have involved the use of physical scale models in contemporary structural design (Addis 2013). Nowadays, form-finding is primarily performed through digital simulations, as a result of the challenges and costs of building reliable scale models that can accurately reproduce the actual properties of the used material. It is more common, particularly for those projects that involve complex geometries and a high degree of prefabrication and automation at the construction site, to produce building elements and components directly from CAD files. Nicolò's vacuum-formed plastic or 3D-printed shells have most likely been used as working models to inform and develop the design of the interiors, and to understand the architectural and spatial qualities of the rooms in a house. In fact, these models were made with clear plastic in order to be able to study and evaluate the shell shape together with the house plan, which was positioned as the base of the model. Such physical models can also be shown to a client to present and discuss preliminary design ideas (Fig. 1.32).

The structural and constructive feasibility of the shell digital model of the villa in Malibu was verified in collaboration with the engineer Ignacio Barandiaran, from the Arup office in San Francisco. The structural and constructive feasibility of such models was also discussed and finalised in collaboration with another engineer, Joe

Hoffmayer, who brought many years of experience to the table, which he had gained as a practitioner in the Los Angeles area; he was therefore able to ensure that the project complied with the local building codes.

Even though the perimeter of the house plan might look like a whiplash inspired by natural forms, this sinuous curve was pragmatically rationalised for construction by simplifying its shape through the use of 53 circular arc segments with different radii. The dining and living areas of the house face onto the private park and the swimming pool in the southwest direction. These two important functions are visually and physically linked to each other through a double-curved entry space. The largest and most significant openings of the house are in fact placed in correspondence to these living areas, and at the southeast end of the building, where the two master bedrooms are located. A gym and a dance room are located at the opposite end of the villa, where they act as buffers between the interior and the outdoor spaces.

A study of the architectural composition of this house would probably end up being a mere academic exercise: the presence of a Binishell, which, in this case, is a double-curved sinuous concrete structure, questions the semantics of conventional building components and systems, such as the façade or the window, as well as the meaning of scale, rhythm and proportions in the elevations. In order to understand the architectural and functional qualities of this project as a whole, it would be more useful to look at the concrete shell structure as a single continuous shelter which—through its curvature and shape—generates the internal spaces and creates the openings that define the natural lighting and visual relationships between the interiors and exteriors of the house. Alastair Gordon synthesised the formal qualities of this structure through an effective visual analogy: "from the air, the Binishell resembles a three-headed turtle shell that's been bleached in the sun", most likely because "there's not a straight line or right angle in sight" (2021: 131).

The constructive logic, which was predominant in Dante Bini's work, is created as a consequence of the plan design in the case of the Malibu villa. Nicolò's response to the functional requirements of the brief defined the architectural and structural features of this Binishell, which also took into account important environmental aspects, such as minimising the use of resources and materials, reducing waste and optimising the impact on the overall health of the planet. As Nicolò Bini stated, "concrete dome structures can therefore provide stronger and safer buildings with lower embodied carbon and lower life-cycle footprints" and, in the last decades, "the challenge with domes has been how to build them quickly, easily and inexpensively and how to make them more effective and desirable architecturally". The Binishell systems that have recently been developed by Nicolò, on the basis of his father's inventions, "provide twice the energy efficiency, with half the carbon footprint, can be built up to three time faster, are safer and more affordable than traditional buildings", and "can also be built using local labour and materials and are highly flexible and dynamic architecturally" (2016). Nicolò argued that the environmental impact of his new Binishell systems, such as the structure built for the villa in Malibu, can be summarised in four simple principles. First, energy efficiency, which is mainly guaranteed by a continuous and insulated building envelope, i.e. the concrete shell structure, that is capable of reducing the number of thermal bridges, whose presence is limited to the position and size of the openings. Second, the overall reduction of building materials, primarily due to the high structural performance of a form-found shell. Third, the minimisation of the waste produced during the construction phase: casting a concrete shell in place by means of a pneumatic membrane eliminates the need to use complex wooden formworks. Fourth, the absence of load-bearing partitions underneath the Binishell leads to great flexibility in the use and readaptation of the internal house layout throughout its life cycle.

It is worth noting that Nicolò's work on this villa in Malibu was certainly facilitated by the presence of a client who was intrigued by a certain kind of unconventional structures and design (Gordon 2021). Robert Downey Jr was particularly fascinated by the spaces presented by Gordon in his book *Spaced Out. Radical Environments of the Psychedelic Sixties* (2008). However, it was surely less straightforward to persuade the local authorities to approve the construction of this Binishell project: there are in fact extremely restrictive structural and seismic requirements in Malibu, as well as specific regulations to ensure a high standard in terms of environmental building performance.

Nicolò submitted a set of documentation drawings that clearly addressed all these aspects, but he primarily convinced the local building officials of the validity of his newly developed Binishell system through the construction of a 1:1 scale mock-up (Fig. 6.17). He built a Minishell-shaped prototype of approximately 90 square metres, at his own expense, on a plot of land owned by the Bini family in Joshua Tree, which is located about 200 kmaway from Los Angeles. In this way, he was also able to show the builder how his System A was supposed to be constructed. Once the building permit had been obtained, Nicolò suggested that the builder should prepare a second and larger mock-up, but of a more complex shape, which should have been closer to the final design of the villa in Malibu, to further try out the construction system. However, this prototype was in fact never made since the builder felt confident enough to proceed directly with the construction of the house, without conducting any further testing.

Nevertheless, Nicolò and his design team conducted an important digital simulation of the whole casting process before proceeding with the site operations. In fact, spraying tonnes of concrete onto a complex shaped pneumatic formwork could have slightly changed the shape of the formwork itself, during the casting, and even have caused the concrete shell to crack. This would not have been a structurally relevant issue, but it would nevertheless have alarmed the client, the builder and the local building officer. In order to limit the number and extension of these small defects, the sequence of the casting phases was first studied digitally, and this sequence was then followed scrupulously by the construction company.

The definition of the construction phases of this villa was likely derived from the experience Nicolò had gained from working with his father Dante. Although, on one hand, a technical evolution of the Binishell system can be observed in this project, as a result of the changes in the legislative framework, on the other, a need to erect a thin shell roof can be noted that no longer had to do with the archetype of the cupola that Dante Bini had originally thought up. In this case, the formal complexity became a further occasion to experiment with and improve the construction process

Fig. 6.17 Nicolò Bini, Minishell-shaped prototype, Joshua Tree, California, August 2014. Construction sequence (NB)

of a concrete shell through the use of a pneumatic membrane formwork. In the same way as for Dante's Binishells, the first site phase foresaw the casting of a concrete edge beam that replicated the amoeba-shaped footprint of the building. Reinforcing rods were bolted to the footing edge beam by means of a threaded rod that had been submerged in the casting, and which had an inward slot onto which the pneumatic formwork was anchored (Figs. 6.18 and 6.19).

Fig. 6.18 Nicolò Bini, villa in Malibu, 2012–2017. Aerial view of the slab on ground and the sinuous edge beam (NB)

Fig. 6.19 Nicolò Bini, villa in Malibu, 2012–2017. Detail of an edge beam (NB)

Fig. 6.20 Nicolò Bini, villa in Malibu, 2012–2017. Aerial view of the inflated pneumatic formwork (NB)

Once the pneumatic formwork had been inflated, the internal pressure was kept constant, starting from the reinforcement positioning stage: this involved placing rods and resorting to the use of bar chairs to space the rods from the membrane. The reinforcement rods were placed in such a way as to form a complex network: the density of such rods was intensified in correspondence to the apexes of the shell and of the five large openings at the edges of the BLOB roof (Figs. 6.20, 6.21, 6.22, 6.23, and 6.24).

The metallic springs that characterised the self-forming reinforcement system of the Binishell patented in 1964 by Dante Bini had disappeared. The layout and pattern of the rods was derived from a precise structural dimensioning in the villa in Malibu: it was no longer the result of an empirical approach that had instead characterised the first Binishell cupolas constructed in Italy, which had likely been inspired by the ribbed roof structure of Pier Luigi Nervi's Palazzetto dello Sport in Rome. Moreover, the cutting of the shell, which was required to create the openings in the Binishells designed by Dante, disappeared almost completely in the construction process of the villa in Malibu: where the large openings were foreseen to visually connect the internal spaces with the garden, the reinforcements were not covered with concrete and were cut off once the cast concrete had hardened. Cuts were only made in this complex concrete shell to create three small ovoid-shaped skylights, which were located in the upper part of the building, or for a small, circular porthole, which was placed on the southwest façade in correspondence to a bathroom.

Fig. 6.21 Nicolò Bini, villa in Malibu, 2012–2017. Internal view of the inflated pneumatic formwork (NB)

Fig. 6.22 Nicolò Bini, villa in Malibu, 2012–2017. Placement of the steel reinforcement and bar chairs (NB)

Fig. 6.23 Nicolò Bini, villa in Malibu, 2012–2017. Placement of the steel reinforcement and bar chairs (NB)

Fig. 6.24 Nicolò Bini, villa in Malibu, 2012–2017. Placement of the steel reinforcement and bar chairs (NB)

The photographs and videos taken in 2015 during the construction of the concrete shell structure show an apparently chaotic set of bar chairs and a tangled mess of rods arranged against the whiteness of the inflated pneumatic formwork, according to a logic that is rooted in a scientific structural calculation that guarantees conformity with the regulations in force in the Los Angeles area: such a conformity was on-the-spot verified by a building officer before the concrete had been cast. The concrete mix, which was prepared by adding fly-ash to Portland Cement, was sprayed on 10 June 2015 (Figs. 6.25 and 6.26). Unlike the original Binishell system, the workers could no longer pour the concrete onto a ground plane, but instead had to spray it onto the inflated formwork using a bucket truck. Moreover, the casting method used for the villa in Malibu did not require the use an external second membrane and surface vibrators to compact the concrete, as the old system conceived by Dante instead needed. The 550 square metres of the 18 cm thick shell were cast in a single day, in eight hours of work: this was a very complex operation, from both the technical and practical points of view, which was in part resolved thanks to the expertise of the builder. In any case, when the site operations had been finished, the client, Robert Downey Jr, remarked: "it was Bini's overall vision that we backed. Mike Grosswendt (All Coast Construction) is another hero of the story: completing the shell exterior, the entirety of the interiors, getting it done within budget, while meeting every required approval of the city and coastal commission. He never balked at the notion of building within a realm devoid of right angles" (Gordon 2021: 132).

In order to allow the cast concrete to harden, the pneumatic formwork was maintained at a constant pressure for four days, under the careful surveillance of Nicolò

Fig. 6.25 Nicolò Bini, villa in Malibu, 2012–2017. Concrete spraying stage (NB)

Fig. 6.26 Nicolò Bini, villa in Malibu, 2012–2017. Concrete spraying stage (NB)

Bini himself. The photographs taken after the formwork had been deflated and removed show images of an unprecedented form of architectural brutalism, which resulted from the intrados of the concrete shell being exposed, hence showing its very smooth but also tattooed surface, full of all those extraordinary imperfections that had been generated during the construction process (Fig. 6.27).[3] At this stage, the appreciation of the internal spatial quality of the shell, which is visible in the abovementioned videos and that is still fully enjoyable because of the absence of the partition walls, is influenced by the imperfections of the intrados. In particular, the layout of the circular imprints left by the reinforcement bar chairs, the pattern of the steel rods, which is occasionally visible in some areas, and the different shades of grey of the concrete surface (Fig. 6.28). These photos are important archival documents because they clearly illustrate a rare and singular aspect of the present-day concept of tectonics related to the production of BLOB concrete architecture (Figs. 6.29 and 6.30). However, the house was handed over to the client with the intrados finished with a layer of plaster: the raw nature of the exposed concrete shell is only visible in a small part of the building, in one of the rooms, to recall the construction technique that had been employed. A layer of insulating material was placed at the extrados of the concrete shell for environmental performance reasons. A light-grey-coloured finishing layer, tending towards white, was placed on top of the insulation to make the surface appear smooth and polished.

[3] See, for example: https://www.youtube.com/watch?v=1kYqSEt8Yug and https://www.youtube.com/watch?v=BzxdKNbKHdk. Accessed 5 November 2022.

Fig. 6.27 Nicolò Bini, villa in Malibu, 2012–2017. View of the raw shell intrados after the deflation of the pneumatic formwork (NB)

Fig. 6.28 Nicolò Bini, villa in Malibu, 2012–2017. View of the imperfections on the raw shell intrados (NB)

Fig. 6.29 Nicolò Bini, villa in Malibu, 2012–2017. Completion of the opening edges (NB)

Fig. 6.30 Nicolò Bini, villa in Malibu, 2012–2017. The finished raw concrete shell (NB)

6.4 The Tectonics of BLOBs

The images of the raw intrados of the concrete shell of the villa in Malibu, taken just after the pneumatic formwork had been removed, open up a theoretical reflection on the current meaning of tectonics in the design and construction of thin concrete shells characterised by a complex amoebic form. This villa also takes on a distinct significance with respect to a contemporary interpretation of a design culture that is rooted in exploring the relationship between architectural form and construction. In fact, this is a rare recent case in which a construction technique—a revamped version of the Binishell system—has not only made the building of a complex geometry possible, it has also been specifically and pragmatically developed to construct this house.

It is sufficient to recall that, already back in 1996, Lynn wrote that "blob construction, it must be acknowledged, is only in its nascent stages of development in contemporary architecture culture" and that "with few exceptions, the recent projects that make use of topological surfaces do so for the development of complex roof forms, and roofs, however programmatically complex, are still in the end just roofs". In line with Lynn's assumption, the villa in Malibu falls into those "many experiments in architecture [that] begin with the problem of the long-span roof, however, because it is there that form, structure and tectonics are so intricately entwined" (1998: 176): and it is in fact according to this same logic that the structural shell designed by Nicolò Bini also becomes an envelope, a façade, an architectural space and a dwelling.

Most designers of BLOB architecture do not generally go beyond developing concepts in the digital domain or making scale prototypes, while Nicolò developed and implemented a technology that allowed him to transform a virtual representation of an exotic idea into a material space that can be sensorially perceived by its occupants. According to Lynn: "Blobs constitute a formal intervention in contemporary discussions of tectonics" because "they promise to seep into those gaps in representation where the particular and the general have been forced to reconcile" (1998: 169). BLOBs have an ability to reveal, in an unequivocable manner, the complex and delicate construction process that generates a specific non-standard space in a given time and in a given place. However, the villa in Malibu, if examined once the building has been completed, in fact rejects the ostentation of the construction process in order to achieve a distinct architectural quality thanks to the effects of natural light on a shell intrados that has been rendered to become smooth and clean.

This is a crucial aspect that helps identify the two design approaches put in place by Dante and Nicolò for two different generations of Binishells: Dante's initial priorities were structural efficiency and construction rationality. Therefore, as discussed in Chaps. 2 and 3, the tectonic qualities of his early concrete cupolas were also related to the degree by which the intrados of the shell was able to visually communicate the construction process. On the other hand, Nicolò prioritises a LeCorbusian idea of the architectural quality of a non-standard space, in which the employed construction process becomes just the most appropriate means of achieving "the play of volumes brought together in light".

Despite this difference, the villa in Malibu should be considered as a kind of architecture manifesto because its idea originated from the long tradition of self-shaping Binishells: its design and construction technique should not be considered—or analysed—as a one-off, impromptu and fortuitous professional occasion, which is an unquestionably common fact for many of the recent BLOB constructions completed throughout the world: BLOB projects are generally recognisable because they are in fact conceived to become architectural icons, whereby their formal uniqueness constitutes their main feature. The villa in Malibu exemplifies what Zeynep Mennan stated about the advent of *non-standard*, which "addresses a provocative challenge to modernist standardisation". According to Mennan, "non-standardisation legitimates the singular, as standardisation legitimated the typical" and "the current revival of the organic inserts itself at the very heart of altering logics of material and industrial production" (2008: 180). This principle has been joined with the entrepreneurial skills of Nicolò Bini with the aim of transforming the experience derived from the Malibu project into a praxis for the large-scale and automated construction of both serial and non-standard architectural artefacts, managed by the *Binishells Inc.* company.

In the last decade, Nicolò has been developing a number of projects and prototypes that are intended as much for the construction of luxury houses and green buildings, as for affordable housing, disaster relief housing and military applications. This architectural research is leading Nicolò Bini to design new construction systems that are able to further reduce the resources required to build concrete shell formworks, including the pneumatic ones. He is currently studying the use of plastic material that can rapidly harden above a pneumatic formwork, thus replacing it and saving on the overall use of energy that is necessary to supply the mechanical inflation device for many hours, while waiting for the concrete to harden. The shapes and architectural spaces that can be designed from this research will certainly define a third generation of Binishell systems, thus further empowering the relationship between form and the art of its assembly, that is, tectonics. From late 2021, Nicolò Bini's research has also begun looking beyond the use of pneumatic formworks. In his latest prototypes, Nicolò has investigated the potentiality of hyperbolic paraboloid structures to act as the reinforcement for concrete ruled surfaces. The aim of this study has been to give shape to a low-cost and rectangular plan house—the contemporary quintessence of the primitive shelter as theorised by Laugier—"you thus pass from the concept of the soap bubble to that of the spider-web".

It is possible that this new form of BLOB architecture, which is closely linked to the invention of new construction systems, will further reinforce the relationship between architectural form and its construction with new paradigms, thereby expanding Lynn's theoretical position, according to which: "Blobs enrich the discourse of tectonics by confounding the terms of tectonic discourse. Blobs cannot be reduced to a typological essence: no two blobs are identical, the form and organisation of any given blob is contextually intensive and therefore dependent on exigent conditions for internal organisation. Most importantly, blobs are simultaneously alien and detached from any place yet capable of melding with their contexts. In any definition of the architectonic, there is an implication of the **arche** as being an ideal global singularity where the **tectonic** involves a particular local identity. For these reasons, blobs promise to open up strategic spaces in tectonic discussions, precisely in the discursive spaces where the particular, the multiple, the contingent is conflated with the global singular" (Lynn 1998: 170).

References

1941. Balloon Houses Designed for Defense Workers Bloom Under Virginia Trees. *Life*, 1 December: 34–35.
1941. Ballyhooed Balloon. *The Architectural Forum*, December: 421.
1947. Airform House is improved for U.S. market, also going overseas. *The Architectural Forum*, July: 15.
1950. The Endless House. *The Architectural Forum*, November: 63–64.
1954. Semitropical Igloo. 'Bubble House' is the new version of a practical cement dwelling for $6,500. *Life*, February 22: 75–76.
1954. Air-formed concrete domes. House Hobe Sound, Florida. *Progressive Architecture*, June: 116–119.
1959. Sculpting with Sprayed Concrete. *The Architectural Forum*, October: 167–168.
1964. *Twentieth Century Engineering*. New York: The Museum of Modern Art.
1964. *Proceedings. World Conference on Shell Structures*. Washington, D.C.: National Academy of Sciences—National Research Council.
1967. Bubble-Building. *The Architectural Forum*, June: 85.
Addis, Bill. 2013. 'Toys that save millions': a history of using physical models in structural design. *The Structural Engineer* 91: 11–27
Adriaenssens, Sigrid, Stefan Devoldere, Laurent Ney and Iwan Strauwen. 2010. *Laurent Ney – Shaping Forces*. Brussels: A+ Publishers.
Alini, Luigi. 2016. *Fabrizio Caròla: opere e progetti 1954-2016*. Napoli: CLEAN.
American Concrete Institute. 2005. *ACI 334.3R-05. Construction of Concrete Shells Using Inflated Forms. Reported by Joint ACI-ASCE Committee 334*. Farmington Hills: American Concrete Institute.
Bailey, Anthony. 1967. The Talk of the Town. Bini Shell. *The New Yorker*, June 3: 28.
Billington, David P. 1983. *The Tower and the Bridge: The New Art of Structural Engineering*. New York: Basics Books.
Bini, Dante. 1966a. *Méthode pour l'érection de bâtiments constitués principalement d'une structure portante en une seule pièce, équipement pour réaliser cette méthode, et bâtiment ainsi obtenu*. Patent FR 1489329. August 9.
Bini, Dante. 1966b. *Method for erecting structures*. Patent US 3,462,521. December 12.
Bini, Dante. 1967. A New Pneumatic Technique for the Construction of Thin Shells. In *Proceedings of the 1st International Colloquium on Pneumatic Structures, International association for shell structures*, ed. Arend M. Haas, 52–55. Stuttgart: University of Stuttgart.

A. Pugnale and A. Bologna, *Architecture Beyond the Cupola*, Mathematics and the Built Environment 7,
https://doi.org/10.1007/978-3-031-26735-2

Bini, Dante. 1968. *Un metodo per l'erezione di edifici e strutture a membrana perfezionato nel complesso dell'armatura estensibile, nonché nella compattazione del materiale.* Patent IT 853736. November 21.

Bini, Dante. 1984. 1500 Buildings Shaped by Air. In *Save Proceedings. International Conference. May 6–9, 1984, Sacramento, California*, ed. O. James Vogl, 187–192. Sacramento: Society of American Value Engineers.

Bini, Dante, 1993. Automation in Construction of Shells, Space, Multi-Storey, Super High-Rise and Extra-Terrestrial Structures. In *Fourth International Conference on Space Structures, Space Structure Research Center 4, Vol. 2, University of Surrey, Guildford, UK*, eds. Gerard A.R. Parke and C.M. Howard, 1526–1536. London: Thomas Telford.

Bini, Dante. 2008. Progettista delle stazioni lunari. In *I leoni della Silicon Valley. Storie geniali di italiani all'estero*, ed. Fiorella Kostoris and Gianfranco Rossi, 119–129. Milano: Guerini e Associati.

Bini, Dante. 2009. *A cavallo di un soffio d'aria. L'Architettura Autoformante*. Milano: Guerini e Associati.

Bini, Dante. 2014. *Building with Air.* London: Bibliotheque McLean.

Bini, Dante, and Stefano Pietrogrande. 1989. Self-Shaping Space Structures. *International Journal of Space Structures* 2: 67-80. doi:https://doi.org/10.1177/026635118900400201.

Bini, Nicolò. 2009. Conclusioni. In *A cavallo di un soffio d'aria. L'Architettura Autoformante*, Dante Bini, 129–136. Milano: Guerini e Associati.

Bini, Nicolò. 2014. Conclusions. In *Building with Air*, Dante Bini, 144–146. London: Bibliotheque McLean.

Bini, Nicolò. 2016. Binishells, a suite of new building Systems designed to address major areas of need and growth in construction. In *Proceedings of the Fourth International Conference on Sustainable Construction Materials and Technologies (SCMT4), Las Vegas, Nevada, USA 7–11 August 2016* (Unpublished, prepared for a presentation at the SCMT4 Conference).

Binishells S.p.A. 1965. *Verfahren zum Errichten von Gebäuden.* Patent DE 1684163. August 10.

Boller, Giulia and Matthias Beckh. 2019. Building with air: Heinz Isler's bubble houses and the use of pneumatic construction techniques. In *Water, doors and buildings: the proceedings of the sixth conference of the Construction History Society*, ed. James W. P. Campbell, Nina Baker, Michael Driver, Michael Heaton, Sabine Kuban, Michael Tutton and David Yeomans, 494–506. Cambridge: Construction History Society.

Bologna, Alberto. 2013. *Pier Luigi Nervi negli Stati Uniti. 1952–1979. Master Builder of the Modern Age.* Firenze: Firenze University Press.

Bologna, Alberto. 2015. "Structural art" in Nervi & Bartoli's Industrial Architecture (1949–1971). In *Space of Production. Projects and essays on rationality, atmosphere, and expression in the industrial building*, ed. Jeannette Kuo, 90–97. Zürich: Park Books.

Bologna, Alberto. 2019. *The resistance of Laugier. The classicism of Murcutt / La resistenza di Laugier. Il classicismo di Murcutt.* Siracusa: LetteraVentidue.

Bologna, Alberto. 2020. OPEN Architecture. Museo di arte contemporanea a Beidaihe, Qinhuangdao. UCCA Dune Art Museum, Beidaihe, Qinhuangdao. *L'industria delle costruzioni. Rivista bimestrale di architettura*, 471: 58-65.

Boullée, Étienne-Louis. 1976. Architecture, Essay on Art. In *Boullée & Visionary Architecture: Including Boullée's 'Architecture, Essay on Art'*, Helen Rosenau, 81–116. London: Academy Editions; New York: Harmony Books.

Buchanan, Peter. 1993. *Renzo Piano Building Workshop. Complete Works, 1.* London: Phaidon.

Carter, Peter. 1974. *Mies van der Rohe at work.* London: Pall Mall Press.

Chilton, John. 2000. *Heinz Isler. The Engineer's Contribution to Contemporary Architecture.* London: Thomas Telford.

Creighton, Thomas H. 1959a. The New Sensualism. *Progressive Architecture*, September: 141–147.

Creighton, Thomas H. 1959b. The New Sensualism II. *Progressive Architecture*, October: 180–187.

Dal Co, Francesco. 2016. *Centre Pompidou. Renzo Piano, Richard Rogers and the Making of a Modern Monument.* New Haven and London: Yale University Press.

Docomomo. 2021. *Save it: La Cupola. Dante Bini, Costa Paradiso, Sardinia, Italy, 1970s.* https://www.docomomo.pt/heritage/save-it-la-cupola. Accessed 2 November 2022.

Draper, Powell, Maria E. Moreyra Garlock and David P. Billington. 2008. Chapel Lomas de Cuernavaca. In *Félix Candela. Engineer, Builder, Structural Artist*, Maria E. Moreyra Garlock and David P. Billington, 128–141. New Haven and London: Yale University Press.

Drexler, Arthur. 1960. *Visionary Architecture. Press Release no. 326200.* New York: The Museum of Modern Art. https://www.moma.org/calendar/exhibitions/2554. Accessed 2 November 2022.

Engel, Heino. 1997 (1st edition: 1967). *Tragsysteme, Structure Systems.* Ostfildern-Ruit: Verlag Gerd Hatje.

Fontana, Lucio. 2014. Epilogue. Michelangelo Antonioni's villa in Sardinia. A magnificent Leopardian concrete moon—abandoned on the steep rocks of Costa Paradiso. In *Building with Air*, Dante Bini, 147–159. London: Bibliotheque McLean.

Foxe, David. 2010. Saarinen's Shell Game: Tensions, Structures, and Sounds at MIT. *Nexus Network Journal. Architecture and Mathematics* 2: 191-211. doi:https://doi.org/10.1007/s00004-010-0027-3.

Gargiani, Roberto, and Alberto Bologna. 2016. *The Rhetoric of Pier Luigi Nervi. Concrete and Ferrocement Forms.* Lausanne: EPFL Press; Oxford and New York: Routledge.

Garlock, Maria E. Moreyra, and David P. Billington. 2008. *Félix Candela. Engineer, Builder, Structural Artist.* New Haven and London: Yale University Press.

Godwin, Michael. 1978. Malvern sports centre – An architect's adventure story. *Concrete* 4: 20-25.

Gordon, Alastair. 2008. *Spaced Out. Radical Environments of the Psychedelic Sixties.* New York: Rizzoli.

Gordon, Alastair. 2021. Back to the Future. *The Purist* 23: 126-141.

Head, Jeffrey. 2011. *No Nails, No Lumber: The Bubble Houses of Wallace Neff*. New York: Princeton Architectural Press.

Herzog, Thomas. 1976, *Pneumatic Structures. A Handbook for the Architect and Engineer.* London: Crosby Lockwood.

Holgate, Alan. 1997. *The Art of Structural Engineering: The Work of Jörg Schlaich and his Team.* Stuttgart and London: Axel Menges.

Howard, Herbert Seymour Jr. 1966. *Structure: an architect's approach.* New York, St. Louis, San Francisco, Toronto, London and Sydney: McGraw-Hill.

Huxtable, Ada Louise. 1960. *Pier Luigi Nervi.* New York: George Braziller.

Isler, Heinz. 1967. Pneumatic Shape for Concrete Shells. In *Proceedings of the 1st International Colloquium on Pneumatic Structures, International association for shell structures*, ed. Arend M. Haas, 50–51. Stuttgart: University of Stuttgart.

Isler, Heinz. 1994. Concrete Shells Derived from Experimental Shapes. *Structural Engineering International* 3: 142-147. doi:https://doi.org/10.2749/101686694780601935.

Kurokawa, Kishō. 1977. *Kishō Kurokawa: Metabolism in Architecture.* London: Studio Vista.

Laugier, Marc-Antoine. 1755. *An essay on architecture; in which its true principles are explained, and invariable rules proposed, for directing the judgment and forming the taste of the gentleman and the architect, with regard to the different kinds of buildings, the embellishment of cities, and the planning gardens. Adorned with a frontispiece, designed by Mr. Wale, and curiously engraven.* London: Osborne and Shipton.

Legault, Réjean. 2021. The making of Habitat 67: A Tense Pas de Deux between Moshe Safdie and August Komendant. *Journal of the Society for the Study of Architecture in Canada / Le Journal de la Société pour l'étude de l'architecture au Canada* 1: 30–50. doi:https://doi.org/10.7202/1082359ar.

Liddell, Ian. 2015. Frei Otto and the development of gridshells. *Case Studies in Structural Engineering* 4: 39-49. doi:https://doi.org/10.1016/j.csse.2015.08.001.

Lynn, Greg. 1998. Blobs, or Why Tectonics Is Square and Topology Is Groovy. In *Folds, Bodies & Blobs. Collected Essays*, Greg Lynn, 169–182. Bruxelles: La lettre volée. First published: Lynn, Greg. 1996. Why Tectonics Is Square and Topology Is Groovy. *ANY: Architecture New York* 14: 58–61.

Marsh, James H. III. 1964. Construction of Thin Shell Structures by The "Lift-Shape" Process. In *Proceedings. World Conference on Shell Structures. October 1–4, 1962. San Francisco California*, 447–452. Washington, D.C.: National Academy of Sciences—National Research Council.

Meninato, Pablo. 2018. *Unexpected Affinities. The History of Type in Architectural Project from Laugier to Duchamp*. New York: Routledge.

Mennan, Zeynep. 2008. The question of non standard form. *METU Journal of the Faculty of Architecture* 2: 171-183.

Mohr, Normand W. 1916. *Combined aeroplane and dirigible balloon*. Patent US 1,064,240. April 16.

Mohr, Normand W. 1927. The City of Saint Francis Idealized. *The Architect and Engineer* 2: 83-84.

Mohr, Normand W. 1933. Thumb Tacks and T-Square. *The Architect and Engineer* 8: 4.

Muntz, Jan Furey. 1989. Bubbles for Defense. In *Wallace Neff, 1895–1982: The Romance of Regional Architecture*, ed. Andrea P.A. Belloli, 69–89. San Marino, CA: Huntington Library.

Muttoni, Aurelio. 2011. *The Art of Structures. Introduction to the Functioning of Structures in Architecture*. Lausanne: EPFL Press.

Muttoni, Aurelio, Franco Lurati and Miguel Fernández Ruiz. 2013. Concrete shells – towards efficient structures: construction of an ellipsoidal concrete shell in Switzerland. *Structural Concrete* 1: 43-50. doi:https://doi.org/10.1002/suco.201200058.

Neff, Wallace. 1941a. *Building construction*. Patent US 2,270,229. April 3.

Neff, Wallace. 1941b. *Building construction*. Patent US 2,335,300. November 25.

Neff, Wallace. 1942. *Method and apparatus for constructing sheel-form* (sic) *structures*. Patent US 2,388,701. July 15.

Neff, Wallace. 1952. *Improved method of erecting shell-form concrete structures*. Patent US 2,892,239. June 19.

Neff, Wallace Jr, and Alson Clark. 1986. *Wallace Neff. Architect of California's Golden Age*. Santa Barbara: Capra Press.

Nervi, Pier Luigi. 1950. *Procedimento di costruzione per la realizzazione di superfici resistenti piane o curve costituite da reticolati di nervature in cemento armato, completate o meno da solette di collegamento tra le nervature*. Patent IT 465636. May 19.

Nervi, Pier Luigi. 1965. *Aesthetics and Technology in Building. The Charles Eliot Norton Lectures, 1961–1962*. Cambridge, MA: Harvard University Press.

Oechslin, Werner. 2002. *Otto Wagner, Adolf Loos, and the Road to Modern Architecture*. Cambridge, UK and New York: Cambridge University Press.

Olivieri, Luigi C. 1975. Bellissima la Casa Cupoletta. *Grazia* 1790, June 15: 26–33.

Otto, Frei. 2005. *Frei Otto Complete Works. Lightweight Construction Natural Design*. Basel and Boston: Birkhäuser.

Otto, Frei, and Bodo Rasch. 1995. *Finding Form. Towards an Architecture of the Minimal*. Stuttgart: Axel Menges.

Peters, Brady. 2007. The Smithsonian Courtyard Enclosure: A Case-Study of Digital Design Processes. In *Expanding Bodies: Art-Cities-Environment: Proceedings of the 27th Annual Conference of the Association for Computer Aided Design in Architecture (ACADIA)*, 74–83. Halifax: ACADIA.

Piegl, Les, and Wayne Tiller. 1997 (1st edition: 1966). *The NURBS Book. 2nd Edition*. Berlin: Springer-Verlag.

Pilz, W.K. 1978. Foreword. In *New South Wales. Dept. of Public Works, Construction of Binishell Reinforced Concrete Domes*, 5. Sydney: New South Wales, Australia. Dept. of Public Works.

Popescu, Mariana, Matthias Rippmann, Andrew Liew, Lex Reiter, Robert J. Flatt, Tom Van Mele and Philippe Block. 2021. Structural design, digital fabrication and construction of the cable-net and knitted formwork of the KnitCandela concrete shell. *Structures* 31: 1287-1299. doi:https://doi.org/10.1016/j.istruc.2020.02.013.

Pfeiffer, Bruce Brooks and Peter Goessel. 2015. *Frank Lloyd Wright*. Köln: Taschen.

Pugnale, Alberto. 2018. Integrating Form, Structure and Acoustics: a Computational Reinterpretation of Frei Otto's Design Method and Vision. *Journal of the International Association for Shell and Spatial Structures* 1:75-86. doi:https://doi.org/10.20898/j.iass.2018.195.901.

Pugnale, Alberto, and Alberto Bologna. 2014. Dante Bini's air structures (1964–1979). From early Italian prototypes to the Australian experience. In *Proceedings of the First Construction History Society Conference. Queen's College, University of Cambridge, 11–12th April 2014*, ed. James W.P. Campbell, Wendy Andrews, Nicholas Bill, Karey Draper, Patrick Fleming and Yiting Pan, 355–365. Cambridge UK: Construction History Society-Short Run Press.

Pugnale, Alberto, and Alberto Bologna. 2015. Dante Bini's "New Architectural Formulae": Construction, Collapse and Demolition of Binishells in Australia 1974–2015. In *Architecture. Institutions and Change. Proceedings of the Society of Architectural Historians Australia and New Zealand Vol. 32*, ed. Paul Hogben and Judith O'Callaghan, 488–499. Sydney: SAHANZ.

Pugnale, Alberto and Alberto Bologna. 2017. Dante Bini's Form-Resistant Binishells. *Nexus Network Journal. Architecture and Mathematics* 3: 681-699. doi: https://doi.org/10.1007/s00004-016-0323-7.

Raun, Christian, Mathias K. Kristensen and Poul Henning Kirkegaard. 2011. Dynamic Double Curvature Mould System. In *Computational Design Modeling. Proceedings of the Design Modeling Symposium Berlin 2011*, eds. Christoph Gengnagel, Axel Kilian, Norbert Palz and Fabian Scheurer, 291–300. Berlin and Heidelberg: Springer. doi:https://doi.org/10.1007/978-3-642-23435-4_33.

Ricci, Giulia. 2016. La cupola. In *Dante Bini. Mechatronics*, Antonio Pennacchio and Giulia Ricci, 131-143. Milano: Postmedia Books.

Rosa, Joseph. 2001. *Folds Blobs + Boxes. Architecture in the Digital Era.* Pittsburgh: The Heinz Architectural Center and Carnegie Museum of Art.

Rudolph, Paul. 1970. *The Architecture of Paul Rudolph.* London: Thames & Hudson.

Salvadori, Mario. 1980. *Why Buildings Stand Up. The Strength of Architecture.* New York and London: WW Norton & Company.

Sandaker, Bjørn Normann, Arne Petter Eggen and Mark Cruvellier. 2011 (1st edition: 1992). *The Structural Basis of Architecture.* London and New York. Routledge.

Schittich, Christian. 2015. Roof and Space. *Detail* 1: 6-14.

Scipio, Louis Albert. 1967. *Structural Design Concepts. Some NASA Contributions.* Washington DC: Technology Utilization Division, National Aeronautics and Space Administration.

Steadman, Philip. 2006. Why are most buildings rectangular? *Architectural Research Quarterly* 2: 119-130. doi:https://doi.org/10.1017/S1359135506000200.

Steadman, Philip. 2015. Architectural Doughnuts: Circular-Plan Buildings, with and without Courtyards. *Nexus Network Journal. Architecture and Mathematics* 3:759-783. doi:https://doi.org/10.1007/s00004-015-0270-8.

Stelsel, Kirk. 2018. Finding Form. *Precast Solutions* 2: 16-22.

Styles, Ross. 1975. *Binishells.* BArch Thesis, The University of New South Wales.

Szczegielniak, Anna. 2019. Circular Plans in Contemporary Housing Architecture. In *IOP Conference Series. Materials Science and Engineering* 471: 1-9. doi:https://doi.org/10.1088/1757-899X/471/7/072025.

Todisco, Leonardo, Giuseppe Sanitate and Giuseppe Lacorte. 2017. Geometry and Proportions of the Traditional Trulli of Alberobello. *Nexus Network Journal. Architecture and Mathematics* 3: 701-721. doi:https://doi.org/10.1007/s00004-016-0326-4.

Torroja, Eduardo. 1958. *The Structures Of Eduardo Torroja: An Autobiography Of Engineering Accomplishment*. New York: Dodge.

Turner, Lloyd. 1986. The shape of things to come. *Concrete International* 1: 18-22.

Viollet-le-Duc, Eugène Emmanuel. 1875. *Histoire de l'habitation humaine depuis les temps préhistoriques jusqu'à nos jours.* Paris: Hetzel.

Vrachliotis, Georg, Joachim Kleinmanns, Martin Kunz and Philip Kurz (eds.). 2017. *Frei Otto. Thinking by Modeling.* Leipzig: Spector Books.

Whitehouse, Franklin. 1967. Architectural Happening at Columbia University: To Blow a Bubble That Won't Burst, Make It of Concrete; CONCRETE SHELL INFLATED IN TEST Dome-Shaped Building Here Was Developed in Italy. *The New York Times*, May 17: 49.

West, Mark. 2017. *The Fabric Formwork Book. Methods for Building New Architectural and Structural Forms in Concrete.* London and New York: Routledge.

Williams, Christopher J. K. 2001. The analytic and numerical definition of the geometry of the British Museum Great Court Roof. In *Mathematics & design 2001*, ed. Mark Burry, Sambit Datta, Anthony Dawson and John Rollo, 434-440. Geelong: Deakin University.

Zorgno, Anna Maria. 1992. Beyond the Cubic Prison. *Rassegna,* 49: 74-83.

Index of Names

A. Pugnale and A. Bologna, *Architecture Beyond the Cupola*, Mathematics and the Built Environment 7,
https://doi.org/10.1007/978-3-031-26735-2

Index of Buildings and Places

A. Pugnale and A. Bologna, *Architecture Beyond the Cupola*,
Mathematics and the Built Environment 7,
https://doi.org/10.1007/978-3-031-26735-2

The manufacturer's authorised representative in the EU is Springer Nature Customer Service Centre GmbH, Europaplatz 3, 69115 Heidelberg, Germany. If you have any concerns regarding our products, please contact ProductSafety@springernature.com

Printed and bound by CPI Group (UK) Ltd, Croydon, CR0 4YY
28/09/2026
02224078-0001